C0-AJY-299

Natural Resources Management Practices

A Primer

Cover Photograph

Pinyon-juniper woodlands in the southwestern United States managed to improve water yield and increase livestock forage production (*left*), as well as to produce wood and protective cover for wildlife (*right*).

Title Page Photograph

Stream-gaging station used to measure water yield.

Natural Resources Management Practices

A Primer

Peter F. Ffolliott

Luis A. Bojorquez-Tapia

Mariano Hernandez-Narvaez

Iowa State University Press / Ames

Peter F. Ffolliott, Ph.D., is a professor at the School of Renewable Natural Resources, College of Agriculture, University of Arizona, Tucson. Formerly a Research Forester for the Rocky Mountain Research Station of the USDA Forest Service, Ffolliott has conducted research throughout the United States and internationally, including Mexico, Central and South America, Africa, the Middle East, and Asia. He is the author or coauthor of more than 450 publications on integrated watershed management, dryland forestry, mensurational and inventory techniques, silviculture, land and wildlife management, erosion and sedimentation processes, multiple-use land and resource management, and ecosystem modeling and simulation.

Luis A. Bojorquez-Tapia, Ph.D., is an investigator, Universidad Nacional Autonoma de Mexico, Mexico City. He specializes in conservation ecology and design of natural protected areas. Other areas of focus include: application of geographic information systems to environmental planning and biodiversity conservation, environmental impact assessment, environmental conflict resolution, and regional land-use planning.

Mariano Hernandez-Narvaez, Ph.D., is a research scientist at the School of Renewable Natural Resources, College of Agriculture, University of Arizona, Tucson, and is affiliated with the U.S. Department of Agriculture–Agricultural Research Service at the Southwest Watershed Research Center in Tucson, Arizona. Mariano Hernandez-Narvaez earned a B.S. in Civil Engineering from the National Polytechnic Institute in Mexico City, an M.S. in hydrology and a Ph.D. in watershed management from the University of Arizona. Dr. Hernandez-Narvaez has been responsible for conducting advanced hydrologic and erosion simulation modeling research and field work throughout the United States and internationally.

Iowa State University Press
2121 South State Avenue, Ames, Iowa 50014

Orders: 1-800-862-6657
Office: 1-515-292-0140
Fax: 1-515-292-3348
Web site: www.isupress.com

∞ Printed on acid-free paper in the United States of America

First edition, 2001

Library of Congress Cataloging-in-Publication Data

Ffolliott, Peter F.
Natural resources management practices : a primer / Peter F. Ffolliott, Luis A. Bojorquez-Tapia, Mariano Hernandez-Narvaez.
p. cm.
Includes bibliographical references and index.
ISBN 0-8138-2541-5
1. Natural resources—Management. I. Bojorquez-Tapia, Luis A. II. Hernandez-Narvaez, Mariano. III. Title.

HC85 .F44 2000
333.7—dc21 00-063145

The last digit is the print number: 9 8 7 6 5 4 3 2 1

Contents

Preface

THIS BOOK has been written to introduce natural resources management practices that are used to achieve societal goals and objectives to students enrolled in college curricula focusing on natural resources management, conservation ecology, and environmental sciences. It was also written as a textbook for students in biological sciences who are interested in natural resources management. The book provides an interdisciplinary orientation to natural resources management practices for students who may subsequently move into more specialized coursework in timber, wildlife, fishery, or watershed management, rangeland conservation, or outdoor recreation but are also interested in acquiring a general knowledge of all aspects of natural resources management practices.

The book also serves as a general reference to planners, technicians, and administrators with responsibilities in coordinating natural resources management and to lay people who deal with professionals having responsibilities for the conservation, sustainable development, and use of natural resources in a particular ecosystem or geographic region.

English units of measurement (e.g., inches, cubic feet, pounds) are used in the book. Exceptions are found where the original relationship or information was presented in the International System of Units, informally called the metric system (e.g., centimeters, cubic meters, grams), and where conversion to corresponding English units is awkward. English to metric conversion factors are provided in Appendix 1.

Natural Resources Management Practices

A Primer

1

Introduction

EFFECTIVE NATURAL resources management and land stewardship must increase in the twenty-first century to meet society's wishes for conservation, sustainable development, and use of natural resources. Multiple-use, ecosystem-based management practices are necessary when considering present patterns of natural resource uses and future uses, which will need to operate at efficient scales. Prerequisite to this need is appreciation of how sound management practices can produce water, forage, timber, and other natural resource commodities and amenities; protect soil and water resources, wilderness values, and scenic attractions; and effectively and efficiently integrate natural resources management practices to meet increasing demands for clean water, open space, and uncluttered landscapes.

DEFINITIONS

Natural resources management is the process of guiding and organizing the use of natural resources on the land to provide desired goods and services without harming soil and water resources or detrimentally affecting the environment. Interrelationships among land use, soil and water resources, and the critical linkages between on-site and off-site benefits are recognized in this concept. *Natural resources management practices* are changes in land use, vegetative cover, natural resources use, and other actions that are taken to satisfy multiple-use, ecosystem-based natural resource management objectives.

RELATIONSHIP OF MULTIPLE-USE MANAGEMENT TO ECOSYSTEM-BASED MANAGEMENT

There is little question that the multiple-use philosophy of natural resources management served society reasonably well into the 1980s. Lands were characterized largely in terms of their capacity to provide commodities and amenities for human use. A role of research was discovering the factors that limited the obtainment of these resources, while a key objective of management practices was reducing or removing limitations. Answers to questions about resources required the identification of optimum yields among desired but often competing uses of the resources.

Multiple-use management as it was practiced into the late twentieth century is not necessarily the best way to approach natural resources management in today's world, however. People began to ask far-reaching questions in the 1990s on how to balance a wide range of potential natural resource uses and values. In responding to these questions the National Research Council (1990) proposed an alternative paradigm for the management of forest lands, one embracing an *ecosystem approach.* Fortunately, the ecosystem approach applies equally well to the management of natural resources on all lands. This approach is the basis of this book.

The ecosystem approach broadens the multiple-use philosophy, creating holistic, ecosystem-based management practices. This approach requires natural resources managers to view landscapes as comprehensive living systems; that is, soils, plants, animals, minerals, climate, water, topography, and all of the processes that link them together. This viewpoint has importance beyond traditional commodity and amenity uses. Natural resources management practices that optimize the production or use of one or only a few natural resources can compromise the balance, value, and functional properties of the whole ecosystem. Furthermore, the production or use of individual resources in itself may lead to a situation of nonsustainability.

Objectives of the ecosystem approach to natural resources management in the twenty-first century must create a compatible balance between the ecological and aesthetic conditions of the landscape and the desired sustainable levels of land use and resource yield. These objectives are similar to the objectives of *area-*

oriented, multiple-use management, where information obtained from *resource-oriented, multiple-use management* is applied by a manager in relation to the suitability of a land area for management (Box 1.1).

For ecosystem management to be successful, natural resources managers must become informed about the conditions, capabilities, and natural resource options for lands. Therefore, it is necessary that the appreciation of natural resources that managers obtain through their professional experiences be shared with the public. It is equally important that natural resources managers consider the values and needs of the public in their decision making, rather than deciding what is good or bad for society based on their personal and technical perspective.

Box 1.1

Integration of Resource- and Area-Oriented, Multiple-Use Management With Ecosystem-Based Management (Ridd 1965, Ffolliott et al. 1995, Brooks et al. 1997)

Resource-oriented, multiple-use management is based on knowing how the management of one natural resource (water, forage, *or* timber) affects its use and the use of other natural resources (water, forage, *and* timber); or how one use of a natural resource (water for agricultural purposes) affects other uses of the same natural resource (water for domestic, industrial, or recreational purposes). Substitutions between and among natural resource uses and associated benefit-cost comparisons of alternative combinations of natural resources use are considered. Resource-oriented, multiple-use management deals with a single natural resource with alternative uses or two or more natural resources with alternative uses for each.

Area-oriented, multiple-use management extends the concepts of resource-oriented, multiple-use management by relating natural resources to each other and the needs and wants of people. Physical, biological, economic, and social factors that relate to development and sustained use of natural resources in a particular area is the basis for area-oriented, multiple-use management. This management approach gathers information from resource-oriented, multiple-use management that describes the potential development and sustained use of natural resources for an area. This information is then related to local, regional, and national supplies and demands for the resources. Objectives of area-oriented, multiple-use management are incorporated into ecosystem-based natural resources management.

REFERENCES

Bowes, M. D., and J. V. Krutilla. 1989. Multiple-use management: The economics of public forestlands. Resources for the Future, Washington, D.C.

Boyce, S. G. 1995. Landscape forestry. John Wiley & Sons, Inc., New York.

Brooks, K. N., P. F. Ffolliott, H. M. Gregersen, and L. F. DeBano. 1997. Hydrology and the management of watersheds. Iowa State University Press, Ames, Iowa.

Ffolliott, P. F., K. N. Brooks, H. M. Gregersen, and A. L. Lundgren. 1995. Dryland forestry: Planning and management. John Wiley & Sons, Inc., New York.

Jensen, M. E., and P. S. Bourgeron, technical coordinators. 1994. Ecosystem management: Principles and applications. Volume II, USDA Forest Service, General Technical Report PNW-GTR-318.

Kaufmann, M. R., R. T. Graham, D. A. Boyce, Jr., W. H. Moir, L. Perry, R. T. Reynolds, R. L. Bassett, P. Mehlhop, C. B. Edminster, W. M. Block, and P. S. Corn. 1994. An ecological basis for ecosystem management. USDA Forest Service, General Technical Report RM-246.

Kessler, W. B., H. Salwasser, C. W. Cartwright, and J. A. Caplan. 1992. New perspectives for sustainable natural resources management. Ecological Applications 2:221–225.

National Research Council. 1990. Forestry research: A mandate for change. National Academy of Science, Washington, D.C.

Ridd, M. K. 1965. Area-oriented multiple use analysis. USDA Forest Service, Research Paper INT-21.

2

Watershed and Water Management Practices

WATERSHED LANDS play critical roles in satisfying human needs. A watershed can contribute forage for livestock and wildlife species, furnish a diversity of primary wood products, and yield water for municipal, agricultural, and industrial developments. Therefore, those concerned with watershed resources need to understand the general principles and basic concepts of watershed management. This understanding will enable the development and implementation of land-use practices that are compatible with water management principles.

DEFINITIONS

A *watershed* is a topographically delineated area that is drained by a stream system. For management planning purposes, a watershed can be considered as a hydrologic-response unit, a physical-biological unit, or as a socioeconomic unit. A *river basin* is defined similarly, but it is larger in scale than a watershed. The Colorado River Basin, Mississippi River Basin, and Amazon River Basin, the largest river basin in the world, comprise the entire land area that is drained by these rivers and their tributaries.

Watershed management is the process of organizing and guiding land and natural resources' use on a watershed to provide desired goods and services without adversely affecting soil and water resources. Part of the concept of watershed management is recognition of the interrelationships among land use, soil, and water and the linkages between upland and downstream areas. *Watershed management practices* are changes in land use, vegetative cover, and other nonstructural and structural actions carried out on a watershed to achieve watershed management objectives.

Water management is concerned with the management of water resources. *Water management practices* focus on developing and conserving water supplies, as well as determining water quality by measuring the physical, chemical, and biological characteristics of a water resource. Other definitions related to watershed and water management practices are presented later in this chapter.

HYDROLOGIC CYCLE

Managers of watershed lands must address specific questions related to land use. These questions include:

- What land-use activities can take place on watersheds without causing undesirable hydrologic effects, such as flash flooding, erosion and sedimentation, and water quality degradation?
- What land-use alternatives may be implemented to change a hydrologic response for a beneficial purpose, such as increasing water yields, improving water quality by reducing sedimentation, and reducing flood damages?

To answer questions such as these, relationships between watershed management practices and the resulting hydrologic responses can be analyzed by studying the hydrologic cycle (Figure 2.1). The hydrologic processes of the biosphere and the effects of vegetation and soil on these processes must be understood for effective watershed management planning. The hydrologic cycle is complex, but it can be simplified as a series of storage and flow components.

Water Budget Concept

The water budget is a concept in which components of the hydrologic cycle are categorized into input, output, and storage. To illustrate this concept for a watershed at a specified time interval:

$$I - O = \Delta S \qquad (2.1)$$

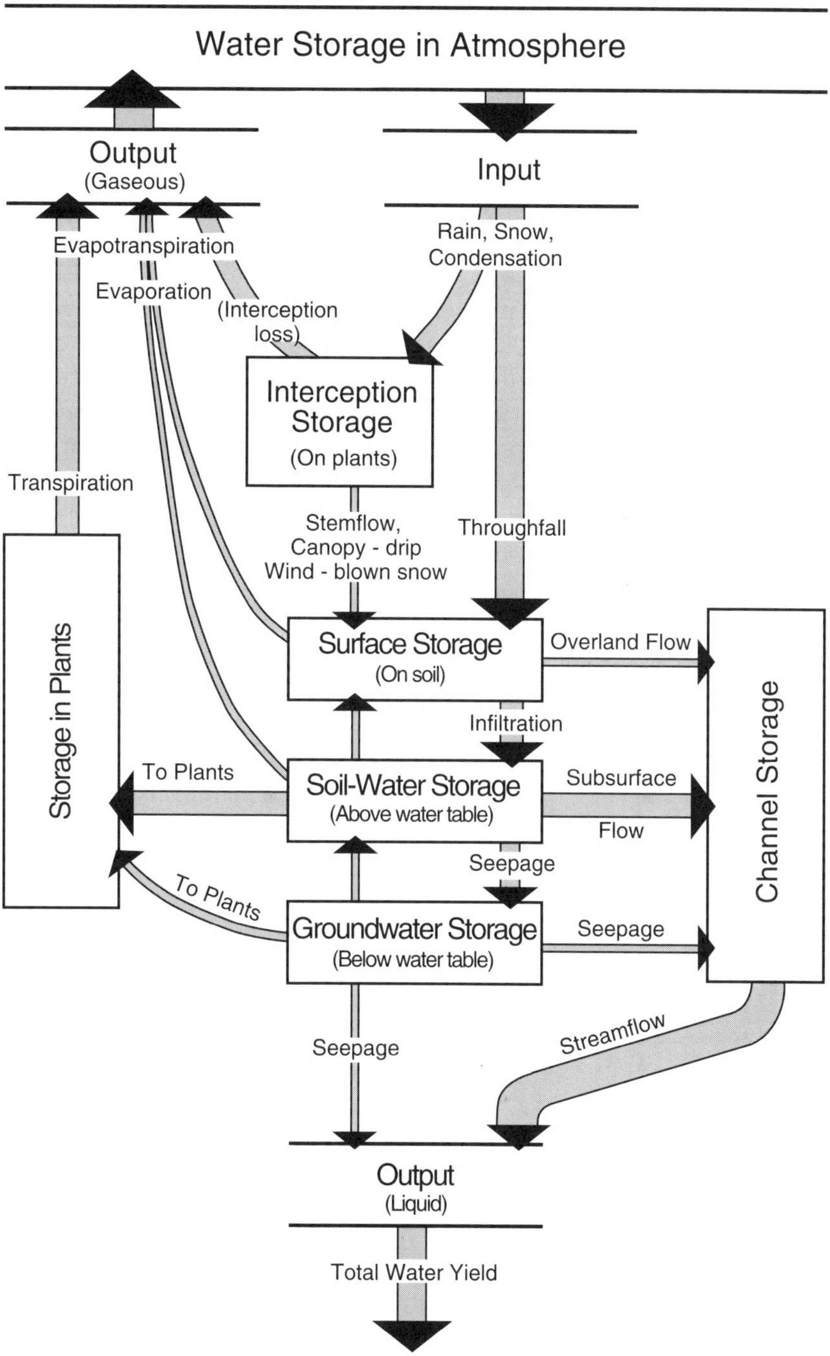

Water Storage in Atmosphere

Output
(Gaseous)

Input

Evapotranspiration

Evaporation

(Interception loss)

Rain, Snow, Condensation

Interception Storage
(On plants)

Transpiration

Stemflow, Canopy - drip Wind - blown snow

Throughfall

Storage in Plants

Surface Storage
(On soil)

Overland Flow

Infiltration

To Plants

Soil-Water Storage
(Above water table)

Subsurface Flow

Seepage

Channel Storage

To Plants

Groundwater Storage
(Below water table)

Seepage

Seepage

Streamflow

Output
(Liquid)

Total Water Yield

Figure 2.1. The hydrologic cycle is a complex system of water storage points (compartments) and the solid, liquid, and gaseous flow of water within and between these storage points (from Anderson et al. 1976).

where I = Inflow of water to the watershed
O = Outflow of water from the watershed
ΔS = Change in storage of the volume of water in the watershed, that is, storage at the end of a period (S_2) minus storage at the beginning of a period (S_1)

The water budget represents an application of the conservation of mass principle to the hydrologic cycle. It is essentially an accounting procedure that quantifies and balances the hydrologic components on a watershed. Coupled with energy, *precipitation* is the primary input to a watershed system. A portion of this precipitation input is *intercepted* by plants and other obstructions and *evaporated* back into the atmosphere, which represents a loss of water from the soil-moisture reserve or the water-flow process. *Infiltration* is the process of water entering the soil surface, while *percolation* is the movement of infiltrated water through a soil body. Water evaporated from soil, plant surfaces, and water bodies and water lost through plant leaves are considered collectively as *evapotranspiration* (ET). ET is the most difficult of all of the water budget components to estimate quantitatively. However, ET and its linkage to soil water storage and the movement of water off of a watershed is one of the hydrologic processes most affected by vegetative manipulations.

That part of the precipitation input that is not consumed through the ET process, but instead runs off a land surface or drains from the soil is the *water-flow component* of the hydrologic cycle. The most direct pathway for precipitation to contribute to streamflow is precipitation input that falls directly into a stream channel or other water body, called *channel interception.* Channel interception causes the initial rise in a streamflow hydrograph, soon after precipitation stops the hydrograph recedes. *Surface runoff,* also referred to as *overland flow,* occurs on impervious land areas or areas on which the rate of precipitation exceeds the infiltration capacity of the soil. Some of the surface runoff is detained by the roughness of the soil surface; nevertheless, it represents a *quickflow response* to a precipitation input. *Subsurface flow,* also called *interflow,* is that part of the precipitation input that infiltrates the soil, but it arrives in the stream channel over a short enough time period to be considered part of the stormflow hydrograph.

A perennial stream, that is, a stream that flows throughout the year, is sustained by *groundwater* between periods of precipitation. Flow from groundwater aquifers does not respond quickly to rainfall because of the long pathways involved and the slow movement of subsurface flow.

One characteristic of stream channels, especially those in dryland environments, is the high *transmission losses* within the channels. When stream channels are dry most of the year, much of the water moving through the systems in a runoff event can infiltrate into the channel. This water is lost from surface streams and ends up as bank storage or percolates into lower soil storage or groundwater systems. As water flows farther downstream, the volume of water in the channel, in the absence of additional inflows, can diminish until there is no longer flow in the channel at some point downstream.

Application of the Water Budget Concept

Application of the water budget concept for the study of the hydrologic behavior of a watershed is relatively straightforward. If all but one component of the hydrologic cycle is measured or estimated accurately, it is possible to solve for the unknown component.

An *annual water budget* is commonly used in a watershed analysis because of the simplifying assumption that annual changes in storage in the watershed are small. Water budget computations can be made beginning and ending with wet months or dry months on the watershed. In either case, the difference in storage between the beginning and end of the annual period is relatively small and is often ignored in calculations. By measuring the precipitation input (P) and streamflow for a year (Q), annual ET can be estimated from:

$$ET = P - Q \tag{2.2}$$

Provided that a reliable measurement of precipitation can be obtained, a second assumption made in forming a water budget is that the total outflow of water from the watershed has been measured. It is assumed that there is no loss of water by deep seepage to underground geological strata and that all of the groundwater flow from the watershed is measured at a gauging station. There are two unknowns in the water budget in such instances, ET and groundwater seepage (L), resulting in the following relationship:

$$ET + L = P - Q \tag{2.3}$$

When it is not appropriate to assume that the change in storage of water in the watershed is small, this change must be estimated. This estimation is difficult, although changes in storage can be estimated by periodic measurements of the soil water content. These measurements are made gravimetrically with neutron attenuation probes or through other methods.

WATERSHED MANAGEMENT PRACTICES

Watershed management includes:

- Practices that maintain a good watershed condition by minimizing adverse impacts on soil and water resources.
- Practices that increase the yields of high-quality water.
- Practices that rehabilitate a degraded watershed from a poor condition into a more productive state.

These watershed management practices are implemented either singularly or in combination, depending on the goals and objectives to be satisfied and the condition of the watersheds on which these are implemented.

Maintaining Watershed Condition

Watershed condition is a term that indicates the health (status) of a watershed in terms of its hydrologic function and soil productivity. *Hydrologic function* relates to the watershed's ability to receive and process precipitation into streamflow. *Soil productivity* reflects the capabilities of a watershed for sustaining plant growth and plant communities, or the natural sequences of plant communities. On a watershed in good condition:

- Precipitation infiltrates into the soil.
- Precipitation does not contribute excessively to erosion, since the resultant overland flow does not dislodge and move soil particles.

- Streamflow response to precipitation is relatively slow.
- Baseflow, or groundwater flow, is sustained between storms.

On a watershed in poor condition:

- A large portion of the precipitation input flows over the soil surface.
- Excessive erosion occurs during precipitation events.
- Streamflow response to precipitation is rapid.
- There is little or no baseflow between storms.

Watershed management practices that maintain a watershed in good condition are those that:

- Sustain high rates of infiltration into the soil.
- Do not contribute to excessive erosion.
- Facilitate a relatively slow streamflow response to precipitation.
- Sustain baseflow between storms on perennial stream systems.

Therefore, watershed management should allow for the following conditions, where possible:

- A protective vegetative cover should be maintained on headwater tributaries, overgrazing of livestock must be precluded, and excessive harvesting of timber must not occur.
- Intensive and concentrated recreational use (picnicking, camping, horseback riding, etc.) of fragile watershed-riparian systems should be avoided.
- Mining of minerals and precious metals, a common point source of soil loss, should be executed with extreme care on hillslopes and eliminated from stream channels.
- Roads and trails established for exploitation or recreational purposes should be dispersed on hillslopes and eliminated from sites adjacent to or in stream channels.
- Engineering measures, such as the construction of check dams (see Chapter 11), should be prescribed only as the first step to control erosion on sites experiencing excessive soil loss. These measures should also be followed by more permanent vegetative measures (seeding, planting of vegetative covers, etc.).

These practices are consistent with the main objective of watershed management: sustaining flows of high-quality water and the availability of other needed resources from watershed lands. When needed, these practices should be incorporated into the planning process to ensure the proper management of watershed lands.

Increasing Water Yields

Vegetative changes that reduce ET generally increase water yields. Evaporative processes generally account for most of the precipitation on a watershed; therefore, the potential to increase water yields by decreasing ET is attractive. ET can be reduced by changes in the structure and composition of vegetation on the watershed.

Yields of high-quality water often increase when:

- Forests are thinned or clear-cut.
- Vegetation on a watershed is converted from deep-rooted plant species to shallow-rooted species.
- Vegetative cover is changed from plant species with high interception capacities to species with lower interception capacities.
- Plant species with high transpiration losses are replaced by plant species with low transpiration losses.

The amount of change in water yields depends on the soil and climatic conditions of the watershed and the percentage of the watershed affected. The largest increases in water yields often result from clear-cutting forests. The length of time that water yields continue to exceed precutting levels is influenced by the type of vegetation that regrows on the site and the rate of this regrowth. High water yield responses are expected in regions with high annual precipitation and deep soils, while responses would be lower in dry climates.

Rehabilitation Activities

Practices to rehabilitate a watershed in a poor condition to a more productive state are the focus of watershed management. These practices include:

- Controlling gullies and soil mass movement by constructing upstream check dams to restrict the movement of eroded soil into the stream channel.
- Establishing tree, shrub, or grass cover on eroded or otherwise degraded sites.

- Limiting timber harvesting, livestock grazing, and road construction on sites experiencing soil loss or on sites that will potentially be subjected to excessive soil loss.

Because severely degraded watersheds are difficult to rehabilitate, initial planning of watershed management practices should incorporate practices that minimize adverse impacts to soil and water resources. By doing so, chain reactions that proceed from an initial loss of soil and water resources to advanced states of degradation in which the overall productivity of a watershed is minimal can be avoided.

WATER MANAGEMENT PRACTICES

The emphasis of water management is generally placed on developing or conserving water supplies. The usefulness of available water supplies to people also depends on its physical, chemical, and biological characteristics. Therefore, both the quantity and quality must be considered in the management of water resources.

Developing Water Supplies

Many methods have been used in developing water supplies in water-deficient regions. Water harvesting is one example. Water harvesting involves the collection and storage of rainfall runoff until the water can be used beneficially. A water harvesting system includes:

- A *catchment area,* the surface of which is often treated to increase runoff efficiency.
- A *storage facility* for collecting water, unless the water is to be utilized immediately, in which case a water spreading system is necessary.
- A *distribution system* when the stored water is to be used later for irrigation purposes.

Water harvesting systems were used by people in the Negev Desert over 4,000 years ago, with applications of this technology continuing to the present (Box 2.1). Some form of water harvesting is often used to sustain livestock and agricultural crop or forest production in dryland environments.

Box 2.1

Water Harvesting Systems: Some Examples

United States

A water harvesting system consisting of a gravity-fed sump, a storage reservoir, 16 catchments, and an irrigation system was constructed on nearly five acres of retired farmland near Tucson, Arizona (Karpiscak et al. 1984). The catchments, treated with NaCl to decrease infiltration, concentrate rainfall runoff around planted tree species and vegetable crops located in untreated areas in the center of the sloping catchments. Excessive runoff flows into a collecting channel and then into the sump. The irrigation system consists of a pipeline connecting the sump and the storage reservoir to a pump, pipelines connecting the pump to the catchments, and drip lines equipped with drip emitters.

Western Australia

To provide water for livestock use and domestic needs, the Public Works Department installed a water harvesting system in a remote region of Australia. The system consists of roaded catchments that were made by clearing, shaping, and contouring (to control length and slope) and then compacting soil in the catchments with pneumatic rollers (Burdass 1975). About 2,500 catchments approximately 2.5 acres in size supply water for livestock use. Additionally, there are 21 catchments totaling 1,750 acres and ranging in size from 30 to 150 acres that furnish water for domestic needs to small towns in the region.

Israel

Researchers have found various methods to increase runoff flowing from land surface into the catchments of water harvesting systems, including land smoothing and compaction, and forming sodic crusts by spraying asphaltic materials (Hillel 1967). Ratios between contributing and receiving (catchment) areas of 3:1 to 6:1 were effective in areas receiving eight to ten inches of annual rainfall. Water received in the planting sites of the catchments supported unirrigated orchards planted by local farmers. The water provided soil moisture amounts that were equivalent to the entire winter rainfall typical of a Mediterranean climate.

Other methods of developing water supplies for a locale include the construction of deep wells, development of springs, and diversion of water from areas of excess supply.

Conserving Water Supplies

Conservation of available water supplies for use at a later date involves a variety of methods to reduce evaporation, transpiration, and seepage losses. Some of these methods pertain to treatments of water, soil, and plant surfaces, while others methods consist of manipulations of vegetative surfaces.

Among the methods of reducing evaporation from small ponds and livestock tanks are covering these water bodies with blocks of wax, plastic, or rubber sheeting or floating blocks of concrete, polystyrene, or other materials. Liquid chemicals that form mono-molecular layers on a water surface (for example, aliphatic alcohols) have been used on larger bodies of water, although their effectiveness can be limited because of wind and deterioration in the sun. Adverse environmental effects on aquatic organisms can restrict the use of evaporative retardants in natural lakes or reservoirs.

Transpiration losses from plants can be reduced by:

- Replacing plant species that have high transpiration rates with species that have lower transpiration rates.
- Removing *phreatophytes,* that is, plants with deep rooting systems that extend into and can extract water from the shallow water tables of streambanks.
- Planting windbreaks of trees or shrubs to reduce desiccating winds.

Earthen canals and reservoirs that are constructed in pervious soils can lose considerable amounts of water through seepage. Methods of reducing seepage losses from these structures include compaction of the soil, treatment of the soil surfaces with sodium salts to break up aggregates, and lining canals and bottoms of reservoirs with various impervious materials.

Water Quality

Usable water is determined not only by the quantity available but also by the water quality conditions. There can be an abundance of water, but its quality is such that it cannot be used safely for irrigation, domestic consumption, or other uses. Therefore, water supplies must be considered in the context of usable water or water that is suitable for a specified use.

The quality of water is affected by natural, geologic, soil, plant, and atmospheric systems and by land-use practices. Water quality characteristics of concern to people can be grouped into physical, chemical, and biological.

Physical Characteristics

Physical characteristics include suspended sediments, turbidity, thermal pollution, dissolved oxygen, biochemical oxygen demand, pH, acidity, and alkalinity. Rainfall events can result in large volumes of surface runoff in relatively short periods of time, promoting erosion and the transportation of sediments. Streams in dryland environments frequently transport relatively higher levels of suspended sediment and exhibit higher turbidity than streams in more humid regions. Surface water and groundwater resources in dryland regions also tend to be more alkaline, with higher pH values than those found in more humid regions, due to typically higher levels of calcium and salts in the water.

Dissolved oxygen and biochemical oxygen demands (an index of the oxygen-demanding properties of biodegradable material in water) are a concern in situations where biodegradable materials enter perennial streams, lakes, and reservoirs. Care is needed to prevent temperatures from increasing in water bodies supporting populations of cold-water fish species. Temperatures of water in dryland environments are usually high.

Chemical Characteristics

Dissolved chemical constituents in surface water and groundwater systems reflect the characteristics of the drainage area. Weathering of rock, physical-biological processes occurring on watershed lands, and atmospheric deposition all affect the chemical compositions of water. Because it is an excellent solvent, the chemical composition of water changes when it comes into contact with rock surfaces and other soil materials. The longer the contact, the greater is the change. On watersheds with little human disturbance, rock and soil substrates generally control ionic concentrations of calcium, magnesium, potassium, and sodium. Nitrogen and phosphorus concentrations are affected by biological activities, although nitrogen, chlorine, and sulfate anions are also added through atmospheric inputs.

High concentrations of dissolved solids (salts) cause some of the severest limitations to the use of available water. In dryland environments, salts tend to concentrate because of high evaporative rates and limited amounts of water. As a result, salinity frequently exceeds 100 parts per million (ppm). People can tolerate salinity levels of 2,500 to 4,000 ppm, at least temporarily. Livestock can tolerate 3,000 ppm. Irrigation is often restricted to salt tolerant plants. It can be necessary to flush salts through the soils so that levels of salt accumulation do not become toxic.

Biological Characteristics

Biological characteristics of water are determined largely by the organisms that impact the use of water for drinking and other forms of human contact. Disease organisms are associated with situations where human and animal waste is treated improperly or the waste is deposited in close proximity to bodies of water.

EFFECTS OF WATERSHED MANAGEMENT PRACTICES ON WATER RESOURCES

Manipulations of vegetation that are part of the management of natural resources can affect long-term productivity of watershed lands. Of particular concern are impacts on the quantity and quality of water that originates from upland watersheds. Therefore, the environmental effects that watershed management practices can have on the hydrology of watershed lands must be recognized.

Environmental Effects

Many watershed lands are subject to grazing by livestock, harvesting of timber, agricultural cultivation, and other forms of human impact. Harvesting of timber or converting from one vegetative type to another can enhance water supplies if the watersheds are managed properly (see below). However, when watersheds are managed improperly, these vegetative manipulations can lead to:

- Excessive surface runoff.
- Increased soil compaction and surface erosion.
- Increased gully erosion and soil mass movement.
- Increased sedimentation in downstream channels.
- Increased export of nutrients from upland watersheds.
- Increased temperature of streams.

The probability for downstream flooding also increases if surface runoff and downstream sedimentation increases, which may result from improper land practices. Soil erosion and the export of nutrients reduce the nutrient storage and the subsequent level of productivity of a watershed. High stream water temperatures and concentrations of nitrate, phosphorus, and other nutrients can adversely affect some aquatic organisms. Introduction of logging residues into stream systems can lead to high biochemical oxygen demand and reduced dissolved oxygen concentration, also adversely affecting aquatic ecosystems.

Detrimental impacts of vegetative manipulations on watershed lands are minimized with properly planned and implemented watershed management practices. For example, successful revegetation of eroded sites could help return watershed lands to their former conditions. *Best Management Practices* (BMPs) should control nonpoint pollution. The BMPs approach involves the identification and implementation of land-use practices to reduce or prevent nonpoint pollution and other environmental problems. In the case of erosion and sedimentation, many BMPs are known for agricultural, forestry, and road-construction activities. BMPs are not well known for some types of chemical pollutants, however.

Water Yield Increases

Opportunities for water yield improvement through vegetative manipulations on watershed lands are related to annual precipitation amounts. For example, the potential for increasing water yields in the southwestern United States is realized only on watersheds with annual precipitation in excess of 15 to 18 inches. The greatest increases in water yields are observed in regions of high precipitation and where precipitation is concentrated in the cooler seasons of the year when ET losses are minimal.

Increases in water yields are generally reduced as the distance between upland watersheds and downstream reservoirs, or other downstream areas where the water is ultimately used, increases (Box 2.2). These reductions in streamflow result from transmission losses in channels, water evaporation, and transpiration by vegetation along the streambank.

Consumptive use of water by phreatophytes along streambanks can be substantial. In the past, salvaging groundwater through eradication of phreatophytes as a way to increase water yields to downstream users has been attempted in parts of the western United States. One problem with these eradication practices is that phreatophytes have environmental values that can be higher than the value of the salvaged groundwater. As a consequence, phreatophyte eradication programs have been curtailed.

Riparian Ecosystems

Riparian ecosystems are the transition areas between aquatic and adjacent terrestrial ecosystems. They are characterized by distinctive soils and vegetative communities that require free or unbounded water. Standing water and running water habitats are commonly found in these ecosystems. Riparian ecosystems are diverse systems that are subjected to a variety of uses.

Box 2.2

Downstream Water Yields: A Diminishing Flow

Increases in water yields attributed to vegetative manipulations on upland watersheds in the Verde River Basin of central Arizona can be small by the time water has traveled 100 miles to downstream points of use. The proportion of the water yield increase reaching downstream areas is relatively small because of transmission losses, evaporation, seepage, and reservoir spills. Simulation of water routing with and without the implementation of vegetation management practices (Brown and Fogel 1987) has indicated that less than half of the streamflow increase is likely to reach consumptive users downstream.

Because riparian ecosystems represent areas of high forage production, they are attractive to livestock. However, in some cases grazing of forage plants by livestock must be curtailed to maintain protective vegetative cover. Riparian sites are often fenced to exclude livestock where excessive grazing occurs. Relocation of water development construction, salting, and herding of livestock can help protect riparian vegetation when fencing is not feasible.

These ecosystems are also unique wildlife habitats, providing an abundance of food and cover, extensive edges between different types of vegetation (ecotones), and a close supply of water. Riparian ecosystems are frequently corridors of migration for wildlife species.

Riparian buffer strips are vegetated bands along stream channels and around water bodies. They are established or left intact to reduce soil erosion and the subsequent delivery of sediments to aquatic ecosystems, maintain streambank stability, and filter excess nutrients that may occur in runoff water (Figure 2.2). These buffer strips also provide shade, shelter, and food for terrestrial wildlife, fish, and other aquatic organisms, while also creating a visually diversified landscape.

SUMMARY

Watershed management practices involve manipulating natural, agricultural, and human resources to achieve specified objectives, taking into account the economic, social, and institutional factors operating in an area. This chapter introduces the subject of watershed management practices and has provided sufficient information so that you should be able to:

- Understand the importance of the hydrologic cycle in studying the relationships between water management and the resulting hydrologic processes.
- Apply a water budget to study the hydrologic behavior of a watershed that has been subjected to a vegetative manipulation.
- Discuss the nature of watershed management practices that maintain a *good watershed condition* by minimizing adverse impacts on the soil and water resources; that increase the yields of high-quality water; and that rehabilitate a degraded watershed from a *poor condition* into a more productive state.

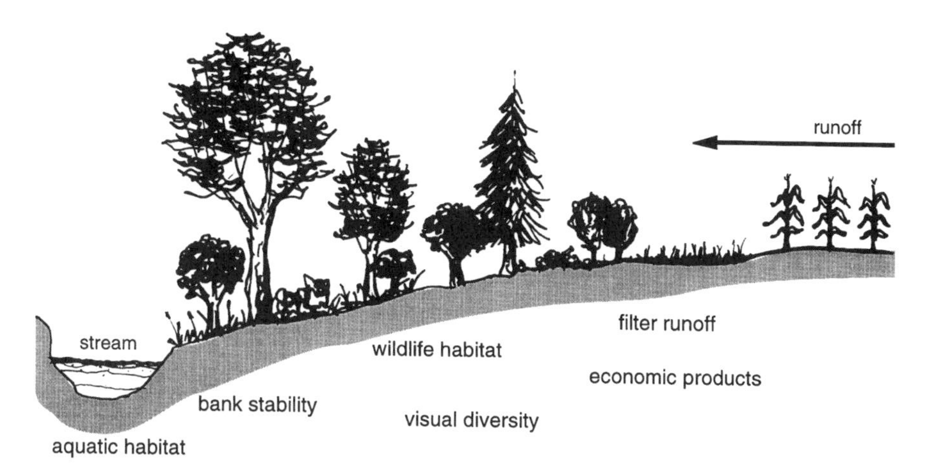

Figure 2.2. Multiple benefits of riparian buffer strips (from USDA National Agroforestry Center 1997).

Both quantity and quality must be considered in the management of water resources. When you have completed this chapter you should appreciate:

- The ways of developing or conserving water resources, including the use of water harvesting methods and methods of reducing evaporation, transpiration, and seepage losses.
- The nature of the physical, dissolved chemical, and biological characteristics of water resources in relation to its usefulness to people.
- How grazing by livestock, agricultural cultivation, harvesting of trees for wood products, and other forms of human impacts affect the availability of water resources.

REFERENCES

Anderson, H. W., M. D. Hoover, and K. G. Reinhart. 1976. Forests and water: Effects of forest management on floods, sedimentation, and water supply. USDA Forest Service, General Technical Report PSW-18.

Branson, F. A., G. F. Gifford, K. G. Renard, and R. F. Hadley. 1981. Rangeland hydrology. Kendall-Hunt Publishing Company, Dubuque, Iowa.

Brooks, K. N., P. F. Ffolliott, H. M. Gregersen, and L. F. DeBano. 1997. Hydrology and the management of watersheds. Iowa State University Press, Ames, Iowa.

Brown, T. C., D. Brown, and D. Binkley. 1993. Laws and programs for controlling nonpoint source pollution in forest areas. Water Resources Bulletin 29:1–13.

Brown, T. C., and M. M. Fogel. 1987. Use of streamflow increases from vegetation management in the Verde River basin. Water Resources Bulletin 23:1149–1160.

Burdass, W. J. 1975. Water harvesting for livestock in western Australia. In: Frasier, G. W., editor. Proceedings of the water harvesting symposium. USDA Agricultural Research Service, ARS-W-22, pp. 8–26.

DeBano, L. P., and L. J. Schmidt. 1990. Potential for enhancing riparian habitats in the southwestern United States with watershed practices. Forest Ecology and Management 33–34:385–403.

Dunne, T., and L. B. Leopold. 1978. Water in environmental planning. W. H. Freeman and Company, San Francisco.

Gordon, N. D., T. A. McMahon, and B. L. Finayson. 1992. Stream hydrology—an introduction for ecologists. John Wiley & Sons, New York.

Heede, B. H. 1980. Stream dynamics: An overview for land managers. USDA Forest Service, General Technical Report RM-72.

Hillel, D. 1967. Runoff inducement in arid lands. USDA Final Technical Report, Project A 10-SWC-36.

Karpiscak, M. M., K. E. Foster, R. L. Rawles, and P. Hataway. 1984. Water harvesting agrisystem: An alternative to groundwater use in Avra Valley area, Arizona. Office of Arid Lands Studies, University of Arizona, Tucson, Arizona.

Lamb, J. C. 1985. Water quality and its control. John Wiley & Sons, Inc., New York.

Newson, M. 1998. Hydrology and the river environment. Oxford University Press, New York.

Renner, H. F., and G. Frasier. 1995. Microcatchment water harvesting for agricultural production: Part I-Physical and technical considerations. Rangelands 17:72–78.

Satterlund, D. R., and P. W. Adams. 1992. Wildland watershed management. John Wiley & Sons, Inc., New York.

Thames, J. L. 1989. Water harvesting. In: FAO. Role of forestry in combating desertification. FAO Conservation Guide 21, pp. 234–252.

USDA National Agroforestry Center. 1997. Riparian buffers for agricultural land. USDA National Agroforestry Center, Agroforestry Notes AF Note 3.

3

Rangeland Management Practices

LIVESTOCK PRODUCTION is a common use of many rangelands and a traditional livelihood for many people. However, rangeland management practices must be compatible with other land uses, including timber production, furnishing wildlife habitats, and agricultural crop production. To understand livestock production and rangeland management practices this chapter discusses the importance of proper rangeland use, grazing management, and rangeland and livestock improvement.

DEFINITIONS

Rangelands are areas on which the native vegetation is predominantly herbaceous and woody plants that are suitable for grazing and browsing by domesticated livestock and wildlife. Rangelands are normally unsuitable for intensive agricultural cultivation. *Rangeland management* is the manipulation of forage and other natural resources on rangelands to provide optimum combinations of goods and services on a sustained basis. Rangeland management is based on ecological principles and deals with careful management of rangelands and range resources.

Forage includes all of the herbaceous and woody plants that are available as food for grazing and browsing animals. *Grazing* is the consumption of forage by livestock or wildlife, while *browsing* is consumption of edible foliage and twigs from woody plants.

A stocking rate is the amount of land allocated to each animal unit for a grazing period. An *animal unit* is one mature (1,000 pound) cow or the equivalent; it is based on an average daily forage consumption of 25 pounds of dry matter. *Proper stocking* refers to placing the correct number of animals on an area to achieve proper use at the end of a specified grazing period. *Proper use* refers to the amount and timing of forage use that can maintain or improve the range condition for the current year's growth. Proper use also includes consideration of how a continuation of the amount and timing of forage use will affect other rangeland resources.

PROPER USE OF RANGELANDS

Rangelands are crucial for livestock production in the western United States. However, even when only livestock grazing is emphasized, impacts of management on other rangeland values (wildlife habitats, watershed protection, etc.) must also be considered. Otherwise, there can be unwanted effects.

Plant Growth and Utilization

Forage plants are the focus of rangeland management. For sustained use of forage plants, the correct number of grazing livestock on a rangeland must be balanced with available forage resources. To achieve this balance and to properly manage grazing on rangelands, an understanding of plant physiology, morphology, and ecology is required. Forage plants are *primary producers;* that is, plants synthesize food from sunlight at rates determined largely by total leaf surface area and water availability.

Livestock are *secondary producers,* since their productivity is determined by the rate at which energy is accumulated by plants and then converted into animal products. Maximum productivity of a rangeland is achieved by maximizing the intakes of forage plants by livestock. However, forage plants are complex organisms. They respond to grazing and a number of interrelated environmental factors and perturbations according to their genetic make up. These relationships must be understood to develop appropriate management strategies of rangeland resources.

Grazing has several effects on the rate of production of forage plants. If grazing is allowed too early in the growing season or too frequently throughout the grazing period, production of forage can be lowered. By defoliating plants, grazing reduces the total leaf area, resulting in lower plant growth rates and thus lower total dry matter production.. However, defoliation can sometimes increase the rate of production for some plants. For maximum forage production, the total leaf area should be maintained close to the optimum for all forage plants throughout the growing season, which is achieved by avoiding severe defoliation and allowing livestock to graze the excess growth as it accumulates.

There is an exception to the grazing result described above. If intense grazing is allowed when a forage plant shifts from vegetative growth to flower production, it can prevent the plant from forming flowering shoots and promote further vegetative growth, including the production of leaves and secondary branches. Consequently, for certain times during extending grazing periods, heavy grazing of forage plants can substantially increase the production of dry matter. However, there is a limit to which any plant can be grazed and still survive.

To determine the proper use of a rangeland, information is needed about the percentage of foliage that can be removed from forage plants without causing overuse and a reduction in range condition. Proper use is determined by clipping forage plants and manipulating grazing levels. Vegetative responses are monitored under different intensities of grazing (defoliation) until thresholds of overuse are identified.

Proper use for a forage species varies with season, climatological conditions, associated plant species, type of livestock, and past grazing history. Therefore, once a proper use factor is established for a targeted forage species, it should be used only as a guideline and should be altered as conditions change. In general, the drier the climate the lower the proper use factor. Examples of proper use factors for forage species vary from 25 to 35 percent in the southwestern United States and from 35 to 45 percent in the high rainfall areas of the short-grass prairielands of the central United States.

Competition is inevitable when different types of livestock (cattle, horses, or sheep) that have similar diets graze a particular rangeland. This competition increases as the total number of livestock on the area increases. However, different types of livestock often have different forage plant preferences. Cattle, horses, and sheep are more adapted to a grass diet, while goats utilize forbs and shrubs when grass production is limited.

Condition and Trend

Rangeland condition is a measure of the health status of a rangeland based on what that rangeland is capable of producing naturally. It is expressed in categories ranging from excellent to very poor. In large part, rangeland condition is an index of whether rangeland management has been applied properly. *Rangeland trend,* on the other hand, is the direction of the change in rangeland condition. Rangeland trend is described as improving, stable, or deteriorating. Both these definitions imply that an assessment of the rangeland is made at a specific point-in-time and in relation to a predetermined standard.

To determine rangeland condition, current plant cover must first be measured. Next, the plant cover necessary for attaining the highest sustained level of forage productivity, at either current site conditions or under conditions of best management practices, is estimated. Rangeland condition is the difference between these values (Box 3.1). When applying the concept of range condition, it is assumed that there is an optimum plant cover for the site. However, the nature of this optimum condition is often unknown, especially on rangelands that have suffered long-term grazing misuse. Therefore, when citing a standard of comparison it should stated whether this standard is known or assumed.

A determination of range condition and trend can also be based on the occurrence and productivity of key forage species that are palatable, nutritious, and abundant under conditions of good rangeland management. However, a careful selection of the key forage species is necessary. A plant should be considered on its own merits and not necessarily on the placement of the forage species in a successional hierarchy. What matters are productivity, persistence, palatability, and cover.

Box 3.1

USDA Forest Service Method of Determining Rangeland Condition

The USDA Forest Service (1969) bases rangeland condition on numerical ratings for vegetal composition, plant production, ground cover, and soil erosion. Vegetal composition, assigned a value from 0 to 60 points, is classified as *desirable, intermediate,* or *least desirable* based on plant lists for reference rangeland sites. Plant production is evaluated by values between 0 and 40 points, with rangelands of highest plant production being given the highest value. Point totals for vegetal composition and plant production are summed to provide a vegetation rating, the maximum rating being 100 points.

Percentage of ground cover and amount of soil erosion are evaluated individually and are each assigned a value from 0 to 50 points. These values are then summed to obtain a soil rating for the rangeland, the maximum rating again being 100 points. Point totals for *either* vegetation or soil are related to rangeland condition as follows:

Points	Condition
81–100	Excellent
61–80	Good
41–60	Fair
21–40	Poor
< 21	Very Poor

It is possible to derive different ratings for vegetation and soil. In these cases, the lower of the vegetation or soil ratings is usually used as the measure of rangeland condition.

GRAZING MANAGEMENT

In properly implemented grazing management practices, livestock are manipulated to meet a designated purpose (for example, to produce meat, milk, or hides) and in a manner that does not adversely impact other land uses. Improper grazing management can lead to rangeland deterioration. The objective of grazing management is to achieve the maximum level of livestock production while simultaneously maintaining or improving the rangeland conditions. Grazing management is based on the following principles:

- Rangelands should be stocked with the proper number of animal units.
- Livestock should be grazed only during the proper season.
- Livestock should be distributed appropriately on a rangeland.
- Rangelands should be stocked with the type of livestock that is best suited to the particular rangeland.

Many rangelands can accommodate more than one type of livestock or wildlife species. Increased production of animal products, which results in a higher economic return to the rangeland, are often attained by grazing more than one type of livestock or combining management of livestock and wildlife species. This form of grazing is referred to as *mixed grazing* or *common use*. For example, inclusion of sheep or goats with cattle, while complicating management procedures, can increase livestock production without adversely impacting rangeland ecosystems. In doing so, a better distribution of livestock may be achieved, resulting in a more uniform use and increased utilization of available forage species.

Stocking Rate

There is a limit to the number of livestock that a rangeland can support. Overstocking should be avoided in most instances. When continued for a long period of time, overstocking will likely lead to rangeland deterioration. On the other hand, understocking reduces the potential of a rangeland to produce livestock products. Therefore, the initial task in properly implementing grazing management practices is to determine the proper stocking rate for a rangeland. If this stocking rate is deemed correct, the second task of management is to maintain the stocking rate. Adjustments in stocking rates are often required throughout long-term grazing periods. These adjustments are made in accordance with seasonal conditions and changes in the production of forage plants as the rangeland responds to the influence of grazing.

Initial stocking rates are calculated for a proposed grazing period by analyzing livestock requirements, dry matter availability, and the nutrient content of forage. Competitive use of forage by wildlife species (if present) must also be considered. The response of a rangeland to an initially calculated stocking rate should be observed for a number of grazing seasons and the stocking rate adjusted so that overgrazing does not occur. This adjusted stocking rate is the *carrying capacity* of the rangeland. A compromise between stocking rates in both good and bad years, weighed in favor of bad years, is generally chosen as the best estimate of stocking rate in most situations.

An initial estimate of carrying capacity can be determined from the following relationship:

$$\frac{(\text{Forage Production})(\text{Forage Utilization})}{\text{Forage Requirement}} \tag{3.1}$$

Estimates of forage production, the amount of forage to be utilized to meet a managerial goal, and the forage requirement to support the type of livestock must be known to solve equation 3.1. A common approximation of the daily dry matter requirements for cattle is two percent of the body weight of the animal. Horses and donkeys have a 50 percent higher requirement. It is frequently assumed that between five and seven sheep are equivalent to one animal unit.

Determining the carrying capacity for a rangeland can be difficult because:

- Steep topography and distribution of water affects animal distribution on the rangeland, requiring that initial stocking rates be adjusted.
- The amount of forage produced in a year varies greatly with changes in year-to-year rainfall patterns.
- Timing of forage removal affects the number of animals that are appropriate.
- Forage competition between grazing and browsing animals is often not quantified.
- Utilization of forage by small mammals and insects needs to be considered.

Determining carrying capacities of rangelands in dryland regions is particularly difficult because of the relatively large year-to-year fluctuations in forage production and prolonged periods of drought have severe effects on forage production. Therefore, stocking rates on these rangelands should be flexible, allowing the number of livestock to be decreased during dry periods and droughts and increased during wet periods and favorable conditions. Unfortunately, livestock producers (ranchers) cannot always have this flexibility in their operations.

Livestock Distribution

It is through control of livestock movement, concentration, or, when necessary, removal that rangeland management is practiced. Proper distribution of livestock on a rangeland to improve overall livestock production can be accomplished by:

- Distribution of salt blocks and mineral licks in different parts of a rangeland to encourage livestock use in areas away from water.
- Development of water in areas where water limits livestock use of available forage resources.
- Herding of livestock, where animals are moved to different forage areas without overgrazing any portion of the rangeland.
- Use of fences.

Controlling the distribution of livestock and the timing and duration of range use by livestock are achieved within the framework of the grazing system used. Implementation of well-planned grazing systems is the basis for attaining a proper distribution of livestock and, therefore, proper rangeland management.

Grazing Systems

Grazing systems should be planned to include: restoring the vigor of forage plants, allowing key forage plants to produce seeds, allowing for more efficient use of a rangeland through appropriate distribution of livestock, and increasing the production of animal products. In general, grazing systems fall into one of four categories: continuous systems, seasonal systems, rotational systems, or balanced rotational systems.

Continuous Systems

As the name implies, in *continuous grazing systems* livestock are kept on one area of land and are allowed to freely move and graze throughout the grazing period. Distribution of livestock is controlled through the placement of salt blocks and mineral licks and the opening and closing of stock tanks, bore-holes, and other water supplies. On rangelands with periodic shortages of water, livestock often adapt a form of rotational grazing, unless they are confined by herding. As a consequence, dry-season and wet-season grazing patterns can be developed separately.

Seasonal Systems

In seasonal grazing systems livestock are purposely confined to one area in the dry season and another area in the wet season. A feature of these systems is that one area of land is grazed at the same time each year. Rangelands that suffer misuse are generally those that are grazed regularly in the wet season. Therefore, a seasonal grazing system is acceptable only when the wet-season grazing area is large in relation to the dry-season grazing area and number of livestock.

Rotational Systems

The total area to be grazed is divided into a number of blocks in *rotational grazing systems.* These blocks are then grazed separately in an order that seems most appropriate to the rangeland manager. Grazing is rationed out in a functional manner, keeping the best forage for the most productive livestock. However, unless grazing is monitored, some of the blocks can be overused and others wasted. When grazing is proportioned equally among the blocks, the result is a balanced rotation.

Balanced Rotational Systems

A period of deferment is applied to each block in *balanced rotational grazing systems,* also referred to as *deferred rotational grazing systems.* Successive grazing periods in a block are deferred so that grazing takes place at different times of the year (Figure 3.1). Each block is grazed for an equal period during a grazing season, which is 12 months on rangelands grazed year round and less on rangelands grazed seasonally. There are many variations of balanced rotational grazing systems, with selection of the grazing cycle among the blocks being determined by the condition of the rangeland, type of livestock, and rangeland management objectives.

Block	Year 1			Year 2			Year 3			Year 4		
	J-A	M-A	S-D	J-A	M-A	S-D	J-A	M-A	S-D	J-A	M-A	S-D
I	G				G				G			
II		G				G				G		
III			G				G				G	
IV				G				G				G

Figure 3.1. A four-block balanced rotational grazing system showing the movement of livestock as each block is grazed (G) for a period of four months. A complete cycle lasts four years and is then repeated (from Pratt and Gwynne 1977).

Other Grazing Systems

Other grazing systems can be designed to satisfy specified rangeland management objectives (Box 3.2). An alternative grazing system should be designed according to the frequency and duration of the grazing periods and, if based on a rotational or balanced rotational system, the length of grazing deferment. The flexibility that can be exercised in management of the grazing systems should also be known.

Box 3.2

Short-Duration Grazing Systems: The Savory Method (Savory 1983)

One other type of grazing system warrants specific mention. *Short-duration grazing systems,* of which the Savory Method is one version, involves overstocking of livestock on small pastures or paddocks, but then moving the livestock quickly before overgrazing occurs. Each pasture is grazed by livestock for a few days, after which the pasture receives several weeks of nonuse. Some rangeland managers believe that stocking rates can be increased by short-duration grazing. Although short-duration grazing systems can result in a more uniform use of forage, other rangeland managers feel that the heavy concentrations of livestock compact the soil surface, which results in reduced rainfall infiltration. This method of grazing livestock emulates the historical patterns of grazing that was prevalent on bison ranges in North America and big game rangelands of Africa.

RANGELAND IMPROVEMENT

Overgrazing can cause a decrease in forage species and available water supplies on a rangeland. When this situation has not progressed too far, efforts to improve range conditions are often initiated. Rangeland improvement methods include removal of undesirable plants to favor the production of more valuable forage species, increasing forage production by seeding the rangeland with forage plants, and water development. Improvement methods must be followed by implementation of rangeland management practices that lead to sustainable resource use.

Removal of Undesirable Plants

Removal of undesirable plants, also referred to as *noxious plants,* to favor the establishment and growth of forage plants can improve rangeland conditions. Undesirable plants include woody and herbaceous species that are unpalatable or poisonous. Removal of these plants can be achieved using mechanical, chemical, or biological methods, or with prescribed fire (see Chapter 10). More than one method of removal will sometimes be necessary when a variety of undesirable plants are present.

Undesirable plants can have varying reactions to different methods of removal. Therefore, to apply a method without considering the physiology of the plant species targeted for removal can be ineffective. Few removal methods result in permanent plant eradication because reencroachment from seeds or resprouting is always possible. Nevertheless, with careful supervision removal of undesirable plants can be achieved, which, when linked to a maintenance program, can increase the amount of available forage.

Seeding of Forage Species

Forage production can be improved on some rangelands by seeding. Seeding of perennial grasses is usually preferred to seeding other plants, with the exception of sites where the rainfall is too low and annual grasses have to be used. Seeding mixtures of grass species is often preferred. These mixtures usually contain annuals in addition to perennials, helping to cover the soil early in the growing season. Mixtures may also contain tufted species to provide a persistent cover and creeping species to quickly cover the soil. Sowing of mixtures can be difficult when the seeds are different in size, shape, and weight. Successful seeding of forage plants should be accompanied with soil treatments and fertilization.

Livestock grazing needs to be excluded on a seeded area during the period of plant establishment. However, light grazing can be permitted in the season following seeding and in some instances can benefit forage species. When a rangeland is grazed under a rotational system, the blocks that are rested should be seeded. If the rangeland is not being grazed in rotation, the area to be seeded should be protected from livestock by fencing or other means. However, protecting seeded areas from wildlife is not always possible.

Water Development

The provision of adequate water for livestock is one of the most important aspects of rangeland management. Water requirements for livestock depend on the type and age of the animals and the ecological conditions of the rangeland (Box 3.3). Livestock can be forced to graze as long as three days between waterings in situations of extreme aridity. Cattle are especially susceptible to water shortages because they require water daily. By grazing as far as ten to 15 miles from their water source, livestock develop an insatiable thirst, with death frequently resulting. Water developments can encourage overgrazing and rangeland degradation if not properly planned and implemented. On rangelands where water supplies are limited, surveying water resources takes precedence over all other considerations in the planning of range improvements.

Box 3.3

Water Requirements for Livestock: Some Considerations

A daily requirement of six to seven gallons of water per animal unit is a general minimum. However, water consumption must rise to improve animal live-weight or milk production. For meat-producing cattle grazed on areas close to water and allowed to drink twice a day, water intake increases to eight to ten gallons per animal. The daily intake can rise to nearly 25 gallons per animal for high-grade, milk-producing cows.

Livestock will consume more water in hot, dry seasons than in cool, wet periods. Nevertheless, it is the maximum seasonal utilization that determines the amount of water that must be provided. For example, if the daily water requirement rises to 5,000 gallons, the water supply must be capable of furnishing this quantity, even when the average for the year is only 1,250 gallons per day.

Methods of water development include:

- Construction of stock tanks—Stock tanks offer a system to trap and hold runoff water that otherwise would be unavailable to livestock.
- Drilling of wells—Wells are a means of increasing available water, but drilling requires knowledge of groundwater aquifers and water quality.
- Development of water harvesting systems—Water harvesting systems are a method of capturing seasonal rainfall and storing the water for use in dry seasons (see Chapter 2).

LIVESTOCK IMPROVEMENT

Livestock that are best suited to the rangeland conditions should be selected for grazing. Desirable traits for the livestock can be heat or cold tolerance, traveling ability, limited water requirements, and efficiency of meat, milk, or wool production. Removal of nonbreeders, extending breeding lives of females, encouraging multiple births, supplemental feeding or flushing in breeding seasons, artificial insemination, and control of breeding seasons all help to improve livestock production.

Through breeding, livestock with desired features (adaptability to climatological conditions, resistance to insects and disease, higher levels of productivity, etc.) are developed. Individual characteristics that are heritable and thus are selected for by breeding include body size, color, growth rate, temperament, mothering ability, and presence or absence of horns. Improved growth rate, one of the more important criterion in a livestock-breeding program, is one of the more difficult of these characteristics to target.

Livestock breeding programs consist of maintaining or refining the livestock type, upgrading the livestock type, crossbreeding, multiple crossbreeding, or rotational breeding. The control of inherited genes, both recessive and dominant characteristics, and varying genetic heritability are prerequisites to the development of a livestock improvement programs.

SUMMARY

The importance of managing rangelands for livestock production should be appreciated because livestock production is a common use of land and a traditional occupation for many people. However, livestock grazing practices should be compatible with forestry, agricultural crop production, and other land uses.

After reading this chapter, you should be able to:

- Explain the concepts of rangeland condition and trend, including what methods are used to measure these indicators, their applications in determining grazing capacities, and their response to different rangeland management practices.
- Understand the important variables in rangeland management, including how to estimate initial stocking rates, the categories of grazing systems, and how these grazing systems are implemented.

- Describe the methods of removing undesirable plants, seeding of forage species, and water development in rangeland improvement programs.
- Recognize situations in which livestock improvement may be warranted.

REFERENCES

Branson, F. A. 1985. Vegetation changes on western rangelands. Range Monograph 2, Society for Range Management, Denver, Colorado.

Heitschmidt, R. K., and J. W. Stuth. 1991. Grazing management: An ecological perspective. Timber Press, Inc., Portland, Oregon.

Holechek, J. L., R. D. Pieper, and C. H. Herbel. 1998. Range management: Principles and practices. Prentice Hall, Upper Saddle River, New Jersey.

Martin, S. C. 1975. Ecology and management of southwestern semidesert grass-shrub ranges: The status of our knowledge. USDA Forest Service, Research Paper RM-156.

Pratt, D. J., and M. D. Gwynne. 1977. Rangeland management and ecology in East Africa. Robert E. Krieger Publishing Company, Huntington, New York.

Savory, A. 1983. The Savory grazing method. Rangelands 5:155–159.

Savory, A., and J. Butterfield. 1999. Holistic management: A new framework for decision-making. Island Press, Covelo, California.

Stoddart, L. A., A. D. Smith, and T. W. Box. 1975. Range management. McGraw-Hill Book Company, New York.

USDA Forest Service. 1969. Range Environment analysis handbook. USDA Forest Service, Intermountain Region, Ogden, Utah.

Vallentine, J. F. 1989. Range development and improvement. Academic Press, San Diego, California.

Vallentine, J. F. 1990. Grazing management. Academic Press, San Diego, California.

Walker, B. H., editor. 1979. Management of semi-arid ecosystems. Elsevier Scientific Publishing Company, New York.

4

Timber Management Practices

FORESTS, WOODLANDS, and plantations are managed when and where it is appropriate to provide food and shelter for livestock and wildlife species, sites for picnic areas and campgrounds, or to provide timber to be processed into lumber, plywood, and paper. Therefore, timber is only one of the multiple uses of these ecosystems. Timber management decisions must consider impacts on other products and uses that can be obtained from forests, woodlands, and plantations.

DEFINITIONS

Timber is wood that is suitable for building, or, more generally, carpentry. *Timber management* is the application of technical information and business methods to operate a forest, woodland, or plantation. *Silviculture* is the art of tending to the forests, woodlands, or plantations to ensure their sustainable development and renewal. *Silvicultural systems* are reproduction methods and intermediate cuttings that sustain and renew forests, woodlands, or plantations until the desired harvest date.

A high forest or low forest is obtained through natural reproduction. A *high forest* is generally of seedling origin and normally develops into a forest of high, closed canopy. A *low forest* is produced by vegetative regeneration, for example, from stump sprouts or root suckers. Planting stock is usually grown in a nursery for artificial reproduction of a forest, woodland, or plantation; this stock can be either bare-rooted or containerized. *Bare-rooted planting stock* is removed from a nursery with its roots freed from the soil in which it grew. *Containerized planting stock* is grown in a receptacle (container) containing soil or other suitable growing media; containerized planting stock is developed either from seed or transplants. Planting stock for some tree species is obtained by vegetative propagation, such as budding, grafting, rooting, and tissue (cell) culture.

Stands are aggregations of trees occupying the same area and are distinguishable from trees found on adjacent areas. Stands are often the primary units of planning and implementing timber management practices. All of the trees in *even-aged stands* are essentially the same age class, although several size classes may be found on a site because of differences in the growth rates of individual trees. Trees in *uneven-aged stands* include two, three, or four age classes and, in most situations, size classes. Each age or size class represents establishment success in a year of natural regeneration.

SILVICULTURAL TREATMENTS

Silvicultural treatments are key to sustaining forests, woodlands, and plantations. A distinction is often made between reproduction methods and silvicultural systems when implementing silvicultural treatments. *Reproductive methods* are the orderly manners by which forests, woodlands, or plantations are established or renewed. Silvicultural systems are more comprehensive in scope, including the applications of reproduction methods and intermediate cuttings throughout the rotational cycle of a forest, woodland, or plantation.

Reproduction Methods

Natural reproduction methods lead to the establishment of natural forests and woodlands. Artificial reproduction methods are used to establish forest plantations of prescribed compositions and densities.

Natural Reproduction Methods

A number of natural reproduction methods are recognized in timber management. These methods and the conditions to which they apply include:

High Forest Methods

- Even-aged stands—Age-class differential is less than 20 years.
  - Clear-cutting method—Removal of an entire stand in one cutting, with reproduction obtained naturally or artificially, if natural regeneration is not possible.
  - Seed tree method—Removal of mature trees in one cutting, except for a specified number of seed trees, which are left singly or in small groups to provide required seed.
  - Shelterwood method—Removal of mature trees in a series of cuttings extending over a relatively short time period. This leads to the establishment of essentially even-aged reproduction beneath the partial shade of seed trees. An area can undergo two or more harvests before the overstory is removed completely.
- Uneven-aged stands—Age-class differential in stands is 20 years or more.
  - Selection method—Removal of the largest or oldest trees, either as single scattered individuals or in small groups, repeated indefinitely. Continuous establishment of reproduction is encouraged and uneven-aged stand structures are maintained.

Low or Coppice Forest Methods

- Coppice method—A type of tree cutting in which renewal depends on vegetative reproduction by stump sprouts or root sprouts.
- Coppice with reserves (standard) method—Production of coppice and high forests on the same site, with trees originating from seeds being "carried through" a much longer time than those of vegetative origin.
- Pollarding—A modification of coppice method where branches of trees are cut to produce sprouts at the top of very high stumps.

In selecting the best reproductive method for implementation, a forester must consider:

- The possible reproduction method that is feasible, that is, germination of seeds or vegetative reproduction from stump sprouts or root suckers.
- The arrangements and timing of the reproduction cuttings.
- The desired future conditions.

Natural reproduction methods that are applied in some of the high forests in the United States are listed in Table 4.1. More than one natural reproduction method can be applied to a forest type in some instances. However, the forester must carefully match the methods to the timber management goals, silvical characteristics of the trees comprising the forest, and the forest structure to be maintained. Tree species in some of forest types can also be reproduced by coppicing.

Artificial Reproduction Methods

Site selection, species selection, site preparation, planting time, planting methods, and spacing and spatial arrangements of the plantings are management considerations when artificial reproduction methods are used in regenerating a forest or establishing a forest plantation.

Table 4.1. Natural reproduction methods for some high forests in the United States

Reproduction method	Forest type
Even-aged stands	
Clearcutting	
West	Douglas-fir, hemlock-Sitka spruce, coastal redwood, ponderosa pine, western larch, Engelman spruce-fir, lodgepole pine
East	Spruce-fir, jack pine, red pine, aspen-birch, southern pines
Seed tree	
West	Ponderosa pine, western larch
East	Jack pine, southern pines
Shelterwood	
West	Douglas-fir, hemlock-Sitka spruce, coastal redwood, ponderosa pine, western larch, Engelman spruce-fir, lodgepole pine
East	Spruce-fir, white pine, jack pine, red pine, northern hardwoods, oak-hickory
Uneven-aged stands	
Selection	
West	Hemlock-Sitka spruce, coastal redwood, ponderosa pine, western larch, Engelmann spruce-fir
East	Spruce-fir, northern hardwoods

Source: Adapted from Young and Giese (1990)

Site Selection. Where to plant is based largely on information obtained from a site reconnaissance. Sites that will lead to the establishment of successful forest plantations in the shortest possible time are desired. The characteristics of the selected site should meet the requirements of the tree species to be planted.

Species Selection. The question of what tree species to plant is often answered at the local level. The aim is selecting species that are suited to the site; will remain healthy throughout the anticipated life of the plantation; attain acceptable growth rates and yield; and meet the objectives of the plantation. Selection of the species to be planted in specific conditions requires reference to planting guides. Planting guides state what species are adaptable to local soils, exposures, and weather patterns and how these species should be planted and maintained. A forester must consider general criteria for selecting tree species when planting guides are not available (Box 4.1).

Box 4.1

General Criteria for Selecting Tree Species to be Planted (Ffolliott and Thames 1983)

Considerations include:

- Native tree species, for which biological and silvicultural knowledge is available, are usually the safest choice. Introduced (exotic) tree species should be used with caution until their local suitability has been demonstrated.
- Tree species with known genetic superiority should be selected.
- Tree species should be selected to meet the following requirements:
  - Ease of obtaining seeds or seedlings for planting.
  - Ease of establishment.
  - Immunity to damaging insects and diseases.
  - Fast growth.
  - Produces useful wood products when this is the purpose of the plantation.
  - Social suitability.

Site Preparation. A site must often be prepared so that planting can proceed without delay once the planting stock arrives from a nursery. Site preparation is undertaken to assure the initial survival and a rapid early growth of planted stock. Dryland environments can demand more intensive and thorough site preparation than is necessary for planting programs in moister climates. Methods of site preparation are:

- Application of approved herbicides.
- Mechanical preparation, such as scalping, that is, removing vegetation and other organic or inorganic material from the area to be planted.
- Prescribed burning.

The site-preparation method selected or combination of methods selected depends on:

- The type of competing vegetation encountered.
- The amount and distribution of rainfall.
- The presence or absence of impermeable layers in the soil.
- A need for protection from desiccating winds, wildfire, and pests, as well as the scale of the planting operation.
- The ultimate monetary value of the trees to be grown is also important, since it determines the amount of money that can be justified for spending on plantation establishment.

Planting Time. The timing of planting often coincides with the onset of the rainy season. This is when bare-rooted stock is planted. The planting operation usually starts as soon as a specified amount of rain has fallen; planting can also be initiated when the soil profile has been wetted to a specified depth by rainfall. A common mistake is planting too soon. If planting is started too late, it can be difficult to complete a large planting program in the scheduled time.

Containerized planting stock has an advantage over bare-rooted planting stock because the soil or other growing media surrounding the roots provides protection during transport to the planting site. It is possible to plant containerized stock in soil that is not sufficiently wet for the successful planting of bare-rooted stock. Therefore, planting time of containerized stock is generally extended beyond the rainy season.

Planting Methods. Hand planting with shovels or planting bars is the easiest and often cheapest method of planting trees. A variety of planting machines are also available for use, although their use is limited on sites with extensive rock cover, and they are expensive to operate. The depth of planting and proper closure of the planting hole can have a greater effect on the initial survival of the trees than on all of the other planting errors. Once they have been planted, the soil surrounding the seedlings must be packed down to avoid the formation of air gaps, which can lead to root desiccation. Unless the seedlings can establish themselves quickly and compensate for transpiration by taking in water through their root systems, they will wilt soon after planting. Therefore, even a single watering immediately after planting can be helpful.

Containerized planting stock is planted with or without the container, depending on whether the container is biodegradable. The stock is planted in holes that are large enough to fit the container or planting plug. Once again, the surrounding soil must be packed down around the seedlings to avoid the formation of air gaps, thus preventing root desiccation. Initial watering of planting stock grown in containers is not as critical as is the case with bare-rooted planting stock because of the protection of the surrounding earthball.

Spacing and Spatial Arrangements. The spacing of seedlings varies with the tree species, site condition, and purpose of the forest or plantation. Where maximum production of total biomass is the management goal, a forester can prescribe closer spacing than that specified in other kinds of forests or plantations. Spatial arrangements of the trees can be rectangular, triangular, or some other geometric pattern. Spatial arrangements of trees in windbreak plantings (see Chapters 5 and 11) are often triangular to provide a more efficient barrier to wind. When mechanical cultivation is to be practiced for maintenance purposes, it is common to plant the seedlings in straight rows and at a width that allows machinery use.

Intermediate Cuttings

Intermediate cuttings either suppress vegetation that impairs growth of residual trees or are designed to achieve prescribed spacing patterns. Intermediate cuttings include:

- Weedings—Cuttings that are made to free crop trees that are not past the sapling stage from competing with vegetation of the same stage, which may currently overtop them or are likely to do so in the future.
- Liberation cuttings—Cuttings made to older, overtopping trees that are competing with immature stands of trees that are not past the sapling stage.
- Thinnings—Cuttings made in immature stands of trees to stimulate the growth rate and increase total production of the stand; a predetermined spacing of the residual trees is the objective of thinning.
- Improvement cuttings—Cuttings to remove trees of undesirable species, form, or vigor in stands of trees that are past sapling stage. This can improve the composition and quality of the stand.
- Salvage cuttings—Cuttings made to remove trees damaged or killed by fire, insects, or diseases. When cuttings are made to prevent spread of insects or diseases from damaged trees, these cuttings are called *sanitation cuttings.*
- Prunings—Cuttings of branchwood on standing trees to increase quality of wood that will ultimately be produced.

Intermediate cuttings are not always economically justified in noncommercial forests and woodlands, even when these cuttings may ultimately increase the volume of the wood grown.

Other Cultural Treatments

Applications of fertilizers and irrigation of trees can be employed in the management of forests and plantations. However, the high investment required in applying these cultural treatments can preclude their large-scale use, although adaptations of an application with a view to increasing productivity can be warranted in some instances.

TIMBER MANAGEMENT

Spacing and arrangement of trees in natural forests and woodlands are typically irregular and random. As a consequence, the costs of management (cultural treatments, harvesting of timber, etc.) can be greater, particularly in forests or woodlands composed of uneven-aged stands, than in forest plantations. In either case, stands are often the primary units for planning and implementing timber management practices.

Species composition and stand structure determine the potential for a stand of trees to meet management goals. A mosaic of stands often develops in a natural forest or woodland, offering a variety of species, stand characteristics, and a range of timber management opportunities. Uniform stands of a single tree species and age represent the common situation in forest plantations.

Species Composition

Tree species are either tolerant or intolerant in even-aged stands. *Tolerance* is a measure of the capacity of a plant species to grow in the shade of other plants. Trees species in uneven-aged stands are generally more tolerant than tree species in even-aged stands, although an overstory of intolerant species related to past disturbances can be found in some situations. Stands composed of a single tree species tend to be even-aged in structure, while stands composed of two or more prominent species generally have an uneven-aged structure.

Stand Structure

Stand structure can be the most important factor in sustaining timber production. An even-aged structure is maintained through harvesting the trees and renewing the forest at one point-in-time. On the other hand, individual trees of a specified age (size) class are harvested to maintain an uneven-aged stand structure. It is sometimes thought that only the largest trees in uneven-aged stands should be harvested. However, sustainable timber production is often compromised when only large trees are removed.

Regulation

In commercial forests, timber is ideally harvested over a continuous time period in about the same or increasing amounts. However, in some instances the age- and size-class distribution of the trees may not allow for harvesting throughout the time period.

Some trees may be too small to harvest, or there can be some years when no trees are suitable for cutting. In these situations, a forester may decide to change the structure of the forest to meet the needs of the timber owner or society, in the case of publicly owned forests, for a sustainable harvest of timber.

A forest that produces a continuous flow of timber is considered to be *regulated.* The amount of timber that is available for harvesting in a specified time period is the *allowable cut.* However, the allowable cut is only a guide and not a unyielding goal that must be attained annually (Box 4.2). Nevertheless, flows of timber harvested must be more-or-less continuous in order to achieve a *sustained yield* of timber.

Attempts to attain regulation include application of area control, volume control, and combinations of these two approaches.

Box 4.2

Regulation: Changing Perspectives (Clutter et al. 1983)

The objective of regulation has changed dramatically in recent years. Attainment of a static and balanced forest may not constitute a reasonable or realistic goal for a forester concerned with obtaining multiple benefits from the forest. In these latter instances, the role of the forester may be management of *imbalanced* stand structures, where the management planning must be coordinated for all of the stands being considered. This situation is often referred to as *forest-level management planning,* in contrast to *stand-level management planning,* which is done independently for each stand.

Area Control

Area control is simple—it means that the volume to be harvested is controlled by the area allocated for timber harvesting. The forest is divided up into a number of smaller areas, each of which is harvested according to a definite cutting schedule so that each year a certain area is available for harvesting. The simplest approach to area control is to divide the area of the forest by the number of years in the rotation age (see below), with the result being the area to be harvested each year. However, the use of area control is not always recommended in the long run because the volume of harvested timber is usually the most critical factor to managers. Furthermore, an equal area harvested each year is likely to produce large variations in the volume of timber harvested annually.

Volume Control

In many instances, volume control, in which predetermined volumes of timber are specified for harvesting from an area, is the recommended method for achieving regulation. Determination of the volume of timber to be harvested to attain volume control is approached through knowledge of the volume of the growing stock, its growth, or both. From these measurements foresters have prepared numerous formula rules and simulation techniques for achieving regulation by volume control. Applications of formula rules, including input requirements and interpretations of the solutions obtained, are found in Clutter et al. (1983), Davis and Johnson (1987), and other references on timber management practices. In recent years formula rules have been mostly replaced by simulation techniques as the basis to obtain a regulation by volume control of a forest, woodland, or plantation.

Simulation techniques are generally based on variable-density yield tables for application to even-aged stand structures and stand table projections for application to uneven-aged stand structures. *Variable-density yield tables* are structured to show the growth and yield of even-aged stands by age, site quality, and stand density descriptors (basal area, number of trees, etc.). The volume of timber to be harvested to regulate these stands is determined through estimates of growth and yields derived by changing the values of these variables (age, site quality, and stand density) to represent future stand conditions.

Stand table projections require development of a *stand table,* a frequency distribution of the number of trees by species and diameter classes on a per unit-area (acre) basis; estimates of diameter growth of trees in a stand at the beginning of a projection period; anticipated removals (losses) by harvesting and natural mortality in the projection period; and estimates of *ingrowth,* that is, trees growing into the smallest diameter class considered for harvest. This information is used to modify the stand table that represents the conditions at the beginning of the projection period to a stand table that represents the conditions at the end of the projection period. For regulation purposes, changes in the two stand tables are then translated into corresponding volume changes by applying volume tables that estimate growth and yield.

Computer simulators based on variable-density yield tables, stand table projections, or combinations of both options have been formulated to help attain regulation (Wiant et al. 1986, Ek et al. 1988). Timber management objectives and the structure of the stands to be regulated determine the best option that should be applied and the computer simulator used.

Rotation Age

The appropriate time to harvest trees in even-aged stands is based primarily on *rotation age.* Foresters use their knowledge of the average annual growth rate of trees comprising even-aged stands to guide them in determining the rotational age and the rotation period of even-aged stands (Box 4.3). *Average annual growth rates* of trees in an even-aged stand is the ratio of the average volume of the trees in the stand to their average age. In general, average annual growth of the trees increases slowly during the initial years of tree establishment, reaches a maximum, and then gradually declines (Figure 4.1). From a biological standpoint, trees should not be allowed to grow beyond the point where average annual growth rates are maximized, which represents the maximum productivity of the stand.

Box 4.3

Determination of Annual Growth Rate: Some Considerations

The volume and age of trees comprising an even-aged stand must be known to determine the annual growth rate at a point-in-time. The average annual growth is determined by dividing the estimated standing volume at a point-in-time by the corresponding age of the stand. However, estimating tree age can be a difficult task in some cases. Annual rings, which are often used to estimate age, are not always formed by trees in dryland environments or humid ecosystems, preventing the use of annual increment counts to estimate age. When this is the case, approximate ages must be estimated from the general size, shape, branching characteristics, and bark condition, but this can only be performed for a few species.

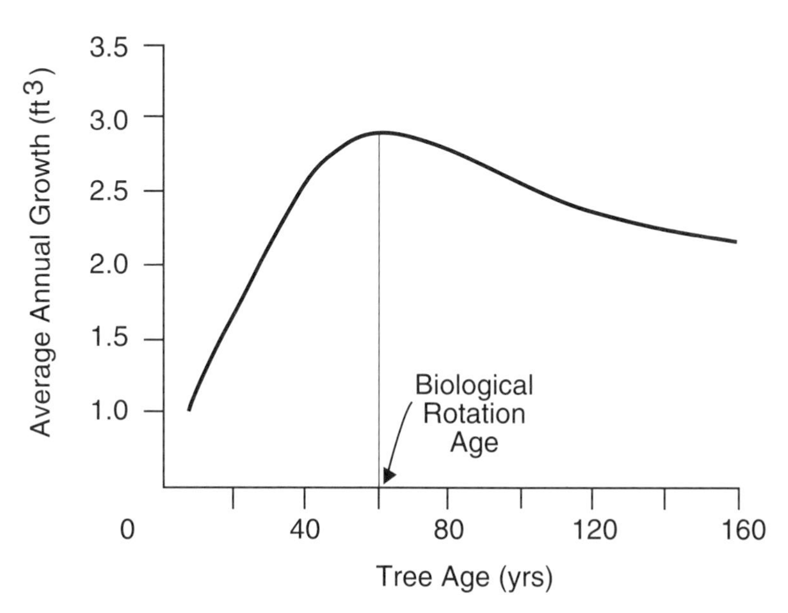

Figure 4.1. The biological rotation age of the trees is the age where average annual growth rate of trees in even-aged stands is maximized (adapted from Young and Giese 1990).

Determining the time to harvest trees in uneven-aged stands is more complicated than with even-aged stands because of the occurrence of trees of different ages and sizes. There are three interrelated factors that must be taken into account when estimating the time to harvest trees being managed in uneven-aged stands:

- The size of trees to be grown, recognizing that trees of varying age classes, and often varying species, occur together.

- The residual volume of trees, referred to as the *growing stock,* that must be maintained to provide adequate growth rates and yields in the future.
- The structure of stands remaining after harvesting that is necessary to provide opportunities for recurring natural regeneration, if required, and orderly growth and development of smaller trees to replace those harvested.

Other biological factors that can influence the decision of when to harvest trees in either even-aged or uneven-aged stands include:

- Pathological factors—Trees can become increasingly susceptible to disease, such as heart-rotting fungi (see Chapter 10), as they increase in age; therefore, they should be harvested before a disease becomes acute.
- Entomological factors—Trees also become susceptible to insect attacks as they get older; therefore, they should be harvested before the insect attacks cause unacceptable mortality.
- Silvicultural factors—Trees tend to produce smaller amounts of viable seed as they get older and become less valuable as seed trees for natural reproduction methods.

When based solely on economics, the best time to harvest timber is when the profit from harvesting, processing, and selling the wood is maximized. Profit is maximized when the returns generated from selling the wood minus costs of harvesting and processing the wood into desired products are the greatest. In most situations this point is reached before the biological rotation age.

Protection

Strategies for the protection of forests, woodlands, and plantations from fire, diseases, and insect infestations are essentially similar. However, the risks associated with protecting forest plantations, where one tree species of a single age class dominates, are greater than in natural forests and woodlands. The frequent mixture of different tree species of different age classes in the latter means that species-specific risks are usually minimal. Nevertheless, repeated occurrences of fire, diseases, and insect infestations in natural forests and woodlands can result in serious losses.

A more detailed discussion of fire prevention and control practices, control practices for insects and fungal pests, and integrated pest management is found in Chapter 10.

HARVESTING OF TIMBER

Harvesting of timber in forests, woodlands, and plantations occurs in two steps:

- Felling and bucking of standing trees before their removal from the forest, woodland, or plantation.
- Skidding, loading, and transporting of the timber to the processing mill.

Felling and Bucking

Felling of trees is accomplished with power saws, axes, or mechanized shears. The direction of the fall must be chosen, the stump height determined, and the undercut and backcut made. A skillful felling crew will consider all of these factors before a tree is actually cut.

Once the trees are lying on the ground, they are cut into marketable segments (logs) prior to being moved to designated loading places, referred to as *landings.* This operation, called *bucking,* can also include removal of the limbs *(limbing)* from the bole. One exception to the typical bucking operation occurs when unmarketable tops are removed and the trees are moved as tree-length logs. Bucking is carried out with power saws, crosscut saws, or mechanized shears.

Skidding, Loading, and Transportation

Logs are moved to a mill to be processed into wood products in two phases:

- Movement from the stump to a landing in the forest. Landings are normally located on flat ground adjacent to a transportation system and are often the center of timber harvesting operations.
- Movement from the landing point to the processing mill.

Operations that are undertaken in both of these phases are more capital intensive than felling and bucking. As heavy-duty equipment is often required, the costs of moving the cut trees to the mill can approach 60 to 75 percent of the total timber harvesting costs.

Skidding

Movement of logs from the stump to a landing is called *skidding* when the movement is carried out by pulling the logs along the ground with racked or wheeled vehicles. When steel cables extending from a stationary power unit are used to move the logs, the skidding operation is often referred to as *yarding*. The general term of skidding is used in this chapter to refer to the movement of the marketable segments from the stump to the landing. At times, the logs are loaded into trucks at the location of felling or after they have been bunched into small piles for loading; skidding is not part of a timber harvesting operation in these situations.

Ground Skidding

Ground skidding involves the pulling the logs along the ground with a mobile power source. Mules and horses were the primary power source through the early 1900s. However, while draft animals are still occasionally used, today tracked or wheeled vehicles, referred to as *skidders,* are the primary source of mobile power for ground skidding.

In ground skidding a cable is hooked to a drawbar on the skidder. At the other end of the cable is a shackle to which three to eight *chokers,* self-tightening nooses of wire rope, are attached. Each choker is placed around the end of a log to be moved to the landing. One to several logs are then pulled flat along the ground. One end of the log can be suspended from an arch to reduce friction and minimize environmental damage. Skidders are often equipped with winches for moving the logs from locations that are inaccessible.

Grapple Skidding

Grapples are hinged, tong-like mechanisms with fine teeth on the inside edge, which are opened and closed mechanically or hydraulically, and can be used in place of chokers. Grapples are lowered over the logs to be moved and then closed. One advantage of grapple skidding is that it is timesaving, also costs are reduced because choker setting and subsequent releasing are eliminated.

Cable Skidding

In cable skidding logs are moved to a stationary power unit, called a *yarder,* located at the landing. Flexible steel cables are extended from the power unit to where they are attached to the logs. The logs are then pulled to the landing by winding the cable onto a drum on the power unit. Ground-lead, high-lead, and skyline are three types of cable skidding systems.

Ground-Lead Skidding Systems. These skidding systems represent the simplest method of cable skidding. Lines from the power unit are hauled into the felling area by hand, draft animal, or tractor where they are attached to the logs to be skidded. The logs are then dragged along the ground to the landing and unfastened, after which the lines are pulled back to the felled trees and the method is repeated.

High-Lead Skidding Systems. These skidding systems are used to overcome the problem of logs striking rocks, stumps, and other obstacles in their path when they are dragged to the landing. Lines are elevated at or near the power unit in high-lead skidding systems, furnishing a lifting effect to the front end of the logs as they are being skidded. The logs are hauled more easily with this lifting action. There are many types of power units. *Spars* are supported by cables between 100 and 150 feet high to elevate the lines. *Carriages* are wheeled vehicles that ride back and forth on the lines, and chokers or grapples can be employed in high-lead skidding.

Skyline Skidding Systems. A cable, called a *skyline,* is suspended between two or more spars, trees, or stumps in these skidding systems. A carriage, with chokers or grapples attached, is moved along the skyline, lifting the logs partly or entirely off of the ground during the skidding operation. In a manner that is similar to high-lead systems, a main line and a rehaul line pull the carriage back and forth along the skyline. Skyline skidding systems are used to suspend logs to protect the soil surface and to move logs across streams, adverse terrain, and other obstacles.

Tightline, slackline, and running skyline are three types of riggings for skyline skidding systems. In *tightline systems* the skyline is anchored so that it cannot be moved up or down. The skyline is lowered with *slackline systems* which permit the chokers or grapples and, therefore, the logs to be attached to the carriage. *Running skyline systems* do not utilize a skyline at all. Instead, the carriage rides the rehaul line, with both the main line and rehaul line used to support the skidding load.

Loading

Once the logs have been skidded to a landing, the next operation consists of loading the logs onto vehicles so the logs can be transported to the mill for processing. At one time, hand loading was practiced for loading small pieces of wood; this method is still used in many small timber harvesting operations. However, because of increased labor costs, power systems of loading have largely replaced hand loading to increase the level of production.

Transportation

Transportation of logs by trucks is widespread. Trucks are available in a wide-range of carrying capacities, with selection of the carrying capacity dependent on the size of the timber harvesting operation, size of the logs to be hauled, hauling distances, topography, and climate. When trucks are employed for hauling, it is of utmost importance that delays in operation are minimized. Staggering the operations in time and space is one approach to reducing delay time. Effective communication will also assist in the scheduling of the operations so that delays are decreased.

Environmental Considerations

People are becoming increasingly aware of the environmental issues related to the harvesting of timber. Ugly residual clear-cuts, poorly stocked forests, and depleted soil resources on the harvested site have become a concern. As a result, environmental constraints have been imposed on felling and bucking operations in recent years through the enforcement of forestry regulations. Governmental regulatory agencies have also established criteria that outline acceptable impacts that can occur on an ecosystem from felling, bucking, and other timber harvesting activities. Included in these guidelines are physiographic limitations on where trees can be felled and bucked; restrictions on the cutting of trees in or near waterways; and acceptable disposal methods for the logging residues that are created by felling and bucking.

SUMMARY

Forests, woodlands, and plantations are managed for sustainable timber production, protection, and amenity purposes. When timber production is emphasized, a forester must realize that timber is only one of the multiple uses obtained. Regardless of the purpose of timber management, these ecosystems must be managed carefully to reduce the depletion of needed growing stock.

After you have read this chapter, you should be able to:

- Discuss the importance of silvicultural treatments, protective strategies, and the time of harvesting for forests, woodlands, and plantations when timber production is the purpose of management.
- Appreciate the relative advantages of bare-rooted and containerized planting stock when artificial reproduction is necessary.
- Explain the meaning of regulation and how forests, woodlands, and plantations are regulated by area control and volume control.
- Describe the sequence of steps followed in the harvesting of timber.
- Discuss the environmental considerations that are related to the harvesting of timber from forests, woodlands, and plantations.

REFERENCES

Barrett, J. W., editor. 1994. Regional silviculture of the United States. John Wiley & Sons, Inc., New York.

Clutter, J. L., J. C. Fortson, L. V. Pienaar, G. H. Brister, and R. L. Bailey. 1983. Timber management: A quantitative approach. John Wiley & Sons, Inc., New York.

Conway, S. 1982. Logging practices: Principles of timber harvesting systems. Miller Freeman Publications, San Francisco.

Daniel, T. W., J. A. Helms, and F. S. Baker. 1979. Principles of silviculture. McGraw-Hill Book Company, New York.

Davis, L. S., and K. N. Johnson. 1987. Forest management. McGraw-Hill Book Company, New York.

Ek, A. R., S. R. Shifley, and T. E. Burk, editors. 1988. Forest growth modelling and prediction. USDA Forest Service, General Technical Report NC-120.

Ffolliott, P. F., and J. L. Thames. 1983. Environmentally sound small-scale forestry projects: Guidelines for planning. CODEL-VITA Publications, Arlington, Virginia.

Ffolliott, P. F., K. N. Brooks, H. M. Gregersen, and A. L. Lundgren. 1995. Dryland forestry: Planning and management. John Wiley & Sons, Inc., New York.

Smith, D. M. 1986. The practice of silviculture. John Wiley & Sons, Inc., New York.

Wiant, H. V., Jr., D. O. Yandle, and W. E. Kidd, editors. 1986. Forestry microcomputer software symposium. Division of Forestry, West Virginal University, Morgantown, Virginia.

Walker, L. C. 1998. The North American forests: The geology, ecology, and silviculture. CRC Press, Boca Raton, Florida.

Young, R. A., and R. L. Giese, editors. 1990. Introduction to forest science. John Wiley & Sons, New York.

5

Agroforestry Practices

THERE ARE a variety of land-use practices that combine the production of agricultural crops, livestock production, forestry, and other production systems to provide ecological stability and sustainable benefits to users of the land. These practices are commonly known as *agroforestry practices.* There is increasing awareness in the fields of natural resources management of the benefits of agroforestry land-use interventions. What is meant by agroforestry in terms of natural resources management is the focus of this chapter.

DEFINITIONS

Agroforestry is a form of land use that sustains or increases total yields of a combination of timber, food crops, and livestock production on the same unit of land (either alternately or simultaneously). Management practices suitable to the social and cultural characteristics of local people and to the ecological constraints and economic condition of the area are applied. *Agroforestry* is also simply referred to as the management of "trees plus any other food crops." *Agroforestry practices* involve land use in which woody plants, including trees, shrubs, and/or vines, are grown for wood production with agricultural crops, with or without livestock.

CLASSIFICATION OF AGROFORESTRY PRACTICES

It is helpful to classify agroforestry practices so that the relative effectiveness of the practices used can be appraised and improvement and future implementation plans can be prepared. Much of the information available on agroforestry is descriptive in nature; as a consequence, agroforestry practices are commonly classified in terms of descriptive criteria.

Structural Basis

The structure of agroforestry practices is often described (initially) by the nature of the components (trees, agricultural crops, livestock, etc.) and then by the temporal and spatial arrangements of the components.

Nature of Components

Three components are generally managed by people in agroforestry practices: trees, shrubs, or vines (forestry); agricultural or forage crops (herbage); and animals (livestock). A profusion of names has emerged in classifying agroforestry practices. The names may consist of any combination of the three components and frequently lead to confusion. However, the following terms are generally used to classify agroforestry practices:

- *Agrisilvicultural practices*—The management of agricultural crops and woody perennials (trees, shrubs, or vines).
- *Silvipastoral practices*—The management of woody perennials and livestock.
- *Agrosilvipastoral practices*—The management of agricultural crops, woody perennials, and livestock.

There are other agroforestry practices that interact ecologically and economically with other land-use production components and fall under the definition of general agroforestry. Among these practices are:

- *Apiculture*—Trees managed mainly for honey production.
- *Aquaforestry practices*—Trees lining fish ponds and the leaves of trees that are used as forage for fish production.

Arrangements of Components

Arrangements of the components of agroforestry practices vary in space and time. Spatial arrangements can include simultaneous placement of all of the components, but the components are spaced in a manner that is supportive rather than competitive. These arrangements include:

- *Random mixtures*—There is little orderly placement of the components, although they are likely to occupy their own ecological niche and are able to coexist.
- *Alternative rows* or *Alternative strips*—Forestry and agricultural crop components are placed in two or more rows or strips. These arrangements are effective in controlling soil erosion and can maintain slope stability when positioned along contours. These spatial arrangements are also referred to as alley or hedgerow croppings.
- *Border tree plantings*—Rows of trees or shrubs are established as windbreaks, live fences, or boundary markers. In addition, these trees and shrubs often provide wood products, fodder for livestock, and green manure (organic fertilizer).

Temporal arrangements are rotational practices in which the components of agroforestry are alternated through time. For example, during a fallow period when crops are not planted on a section of land, livestock can be allowed to graze. Common temporal arrangements are:

- *Coincident temporal arrangements*—Each agroforestry component occupies the same piece of land.
- *Concomitant temporal arrangements*—All agroforestry components occur at the same time and are in a side-by-side arrangement.
- *Overlapping temporal arrangements*—One agroforestry component extends over or covers part of another agroforestry component.
- *Sequential temporal arrangements*—One agroforestry component follows another agroforestry component in sequence (series).

Functional Basis

Agroforestry practices are also grouped according to functional categories, that is, productive and protective functions. Agroforestry practices with *productive functions* are those producing food, energy, shelter, building materials, and in some instances cash income to the landowner. *Protective functions* of agroforestry practices contribute to these productive functions by sustaining and, at times, augmenting production functions. This may include improvement of microclimates and retention of soil and water resources. Most agroforestry practices have both productive and protective functions.

Ecological and Socioeconomic Settings

Descriptions of many agroforestry practices occurring in specific ecological settings or geographical regions are found in Nair (1993), MacDicken and Vergara (1990), Rietveld (1995), and Gordon and Newman (1997). Intercropping of mesquite (*Prosopis juliflora)* and barley by the Native Americans in the southwestern United States and northern Mexico is one example. The usefulness of these descriptions in the classification of agroforestry is limited, however. Several agroforestry practices may be used in a particular ecological region. The functional purpose of an agroforestry practice also differs with the environmental condition and ecological stability of the region. Where basic needs of people are not being met, emphasis will likely be placed on production. Emphasis can be placed on protection where sustainability of environmental quality is paramount.

Agroforestry practices are also grouped by socioeconomic categories. A *subsistence, commercial,* or *intermediate* status will be assigned to a practice depending on whether the practice outputs are used to satisfy the basic needs of people, or if they are made available for sale in a marketplace, or a combination of the two. One drawback to this classification scheme is that boundaries of the groupings also change in time, reflecting changing economic conditions in an area. As a consequence, socioeconomic criteria alone are not usually adopted as a basis for classification.

OVERVIEW OF AGROFORESTRY PRACTICES

A multitude of agroforestry practices, designs, and variations in implementation are found in the United States and throughout the world. Among the more common of these agroforestry practices are windbreak plantings, silvipastoralism, alley cropping, management of riparian buffer strips, forest farming practices, home gardening, and growing multiple-purpose trees.

Windbreaks

Windbreaks, also called shelterbelts, living fences, or timberbelts, consist of linear rows of trees and/or shrubs planted at right angles to the prevailing wind pattern. Windbreaks can:

- Reduce soil loss caused by wind erosion.
- Enhance agricultural crop production.
- Protect livestock from extreme heat or cold.
- Protect roads from dangerous crosswinds and blowing snow.
- Provide buffers for the urban-wildland interface, or furnish protection and buffers within urban communities.
- Provide timber.
- Create travel lanes and provide habitat for wildlife.
- Serve as a living fence.

The number of rows of trees or shrubs planted in a windbreak depends on the intended purpose of the windbreak. The landscape characteristics, prevailing wind patterns, and willingness of the landowner to provide land for the windbreak are also important considerations. Three to seven rows of trees, shrubs, or combinations of both are common. Barrier densities of 40 to 60 percent in vegetation provide the greatest protection to downwind areas. To effectively protect downwind areas, windbreak plantings should generally extend out a distance ten to 30 times the height of the tallest trees in the windbreak (Figure 5.1). Spacing between windbreak trees or shrubs should be less than or equal to this distance when multiple windbreaks are planted.

Other criteria for designing windbreak plantings and other considerations for the management of the windbreaks are presented in Chapter 11.

Silvipastoralism

Silvipastoral practices combine timber management with forage and livestock production. Trees provide a microclimate favorable for growing forage and furnish shade and shelter for livestock.

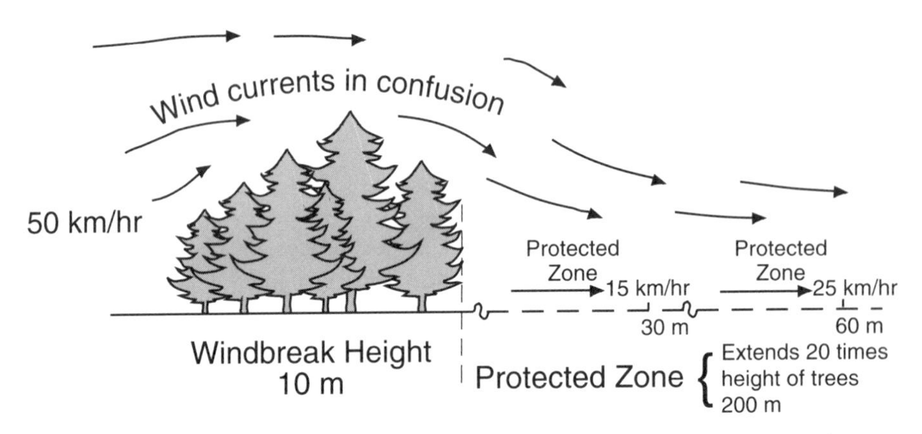

Figure 5.1. Effects of a windbreak planting on wind velocity (from Forbes 1961).

The shelterwood method (see Chapter 4) is one management approach that is often used in establishing a silvipastoral system in forest communities of the western United States. Densities of the tree overstories are reduced in a series of cuttings, increasing light penetration and allowing the establishment of an understory forage crop for livestock consumption.

Trees are also planted on existing rangelands or pastures to create a silvipastorial management system. Spacing of the planted trees is prescribed to balance both wood and forage production: the greater the spacing, the greater the production of forage plants, and vice versa. Protecting the trees from livestock damage is vital in either case.

Deer, elk, or antelope often compete with cattle, sheep, or horses for forage resources. Only by reducing the number of indigenous herbivores, livestock, or both can rangelands that have been degraded by this competition for forage be restored. Therefore, balancing silvipastorial practices with good rangeland management practices (see Chapter 3) is necessary.

Alley Cropping

Alley cropping schemes are formed when trees are grown in either a single row or multiple rows and are spaced far enough apart to form alleys in which agricultural or horticultural crops can be planted. High-valued hardwood species (e.g., oak, walnut, or ash), fruit trees (e.g., apple, peach, or orange), or fast-growing, short-rotation tree species (hybrid poplars) are examples of species that are grown in alley cropping systems. A landowner waiting for a financial return on long-term investment in tree-related crops may suffer a loss of income. However, this loss is partially offset by the income obtained from harvesting the annual agricultural crops grown in the alleys or from the fruits and nuts produced from the trees. Alley cropping is one way to convert marginal agricultural lands to forest or woodland. A benefit to this strategy is that in the initial years of the conversion, annual crops continue to generate an income.

Tree species selected for alley cropping systems should be compatible with the local site conditions, satisfy the landowner objectives, and yield a marketable product. Hardwood species like walnut, oak, ash, and pecan are favored in alley cropping systems because they can potentially provide high-value lumber, veneer logs, and other wood products. On sloping terrains, rows of trees should be planted along the contours to reduce the rate of soil erosion. Trees should also be spaced wide enough apart to permit machinery to maneuver around the trees.

Riparian Buffer Strips

Riparian buffer strips (see Chapter 2), which consists of perennial vegetation alongside streams, lakes and ponds, wetlands, and drainage ditches, can:

- Stabilize streambanks.
- Protect floodplains.
- Reduce nonpoint source pollution.
- Enhance aquatic and terrestrial habitats.
- Improve landscape appearance.
- Serve as protective barriers against agricultural, urban, or industrial land-use practices on adjacent lands.
- Process water from field drainage systems.
- Provide harvestable and marketable products, and in some situations function as a windbreak.

Riparian buffer strips can be generally classified into one of the following categories:

- Strips of perennial trees, shrubs, or herbaceous species planted between the upland watershed land and a stream, lake or pond, wetland, or drainage ditch.
- Strips of naturally regenerated trees, shrubs, and herbaceous plants growing along the edges of a waterway or water body.

Riparian buffer strips are planned and managed to perform a desired function. To achieve the planned function of a buffer strip, the appropriate plant species composition, planting width, and maintenance activities are selected. When control of the sediment carried in surface runoff is a primary purpose of a buffer strip, a grass cover alone can be superior to other plantings. Combinations of grasses, trees, and shrubs function best when control of agriculture-related chemical (fertilizer) runoff, streambank stabilization, or protection and creation of wildlife or fish habitats are the primary objectives (Figure 2.2). A diversity of benefits are possible with a multi-strata buffer, which consists of strips of trees, shrubs, and herbaceous plants paralleling the drainage. Design and maintenance strategies to achieve the benefits of riparian buffer strips are determined by:

- The objectives of the buffer strips (production, protection, or both).
- The landscape physiology, drainage pattern, and other site characteristics.
- The presence of livestock grazing, row cropping, protection of waterways, or other land-use practices.
- Maintenance requirements (low intensity, high intensity).

Forest Farming Practices

Forest farming practices combine agriculture with forestry to produce some kind of specialty product. An established forest, woodland, or plantation is modified as necessary to produce an agricultural crop in addition to existing timber production. A key factor to successful forest farming is a specifically planned production system that is intentionally created and intensively managed for a stated purpose. Removal of some trees may be required to create the site conditions necessary for growing agricultural crops in the forest understory. Examples of forest farming practices are:

- Maple syrup production.
- Growing of medicinal plants, for example, ginseng.
- Harvesting grasses, branches and tree burls, pine cones and seedpods, evergreen cuttings, and other craft materials.
- Mushroom production, for example, American matsutake (*Tricholoma magnivelare)* in the Oregon Cascades.
- Collecting native fruits and nuts.

Home Gardens

Home gardens can include a diversity of trees, shrubs, vines, and herbaceous plants that yield vegetable crops, fruits and nuts, forage and fodder, wood resources, and flowers. Production from home gardens may be for home use or for sale in a marketplace. Garden trees also provide shade for people, their homes, or their livestock and protect other crops. Home gardens are planted in blocks, rows (alley cropping), or random arrangements and can include gardens grown on an adjacent homestead or common community land.

Plant products are obtained from the trees at the upper levels (stratum) of the gardens; from herbaceous plants growing near, on top of, or in the ground; and occasionally from intermediate shrub or vine layers. A variety of wood products are obtained from garden tree levels. Most of these are for household use (e.g., fuel, poles or posts, novelty items, or cottage industries). Cereals, legumes, leaf or root vegetables, forage, cut or potted flowers, and other crops are grown in varying combinations in the herbaceous layers. Plant products of intermediate layers, which are not always present in home gardens, include shrubs of commercial value, grapes, or small fruit-bearing trees. Seeds of trees, shrubs, vines, and herbaceous plants are occasionally collected for future use or commercial distribution.

Animal products from home gardens include the animals themselves (e.g., goats, pigs, or poultry), and products that are derived from animals (e.g., milk, cheese, or eggs). The animal component of home gardens can sometimes roam freely on the landscape. However, it may become necessary to separate the animal component from the plant component by live- or barbed-wire fencing in order to protect the growing plants.

Multiple-Purpose Trees

Multiple-purpose trees are themselves an agroforestry system. The wood is often used for fuel or fence posts; the leaves and small branches furnish livestock fodder; and the fruit and nuts are food for people. This form of agroforestry is illustrated by the tree species listed in Box 5.1. Other potential multiple-purpose tree species that are suitable for specific site conditions are described in MacDicken and Vergara (1990), Nair (1993), and Ffolliott et al. (1995). Efforts to genetically improve multiple-purpose trees have centered on fast-growing, nitrogen-fixing species.

Box 5.1

Multiple-Purpose Tree Species: Some Examples (Gordon and Newman 1997)

There are many multiple-purpose tree species in the United States. In addition to their value for wood, various tree species have been and continue to be utilized for a variety of uses. Maples (*Acer* spp.) are tapped for sugar. Fruits are gathered from walnuts (*Juglans* spp.), oaks (*Quercus* spp.), beeches (*Fagus* spp.), pines (*Pinus* spp.), mesquites (*Prosopis* spp.), and the Rosaceae family, to name a few. Ashes (*Fraxinus* spp.) are lopped for fodder.

Mesquites are multiple-purpose trees of debatable value. Shrub-form mesquites have invaded many of the rangelands and pastures in the southwestern United States, reducing the production of valuable and limited forage species. However, these legumes can also be a source of nitrogen and carbon to semi-desert rangelands deficient in these chemicals. Its wood is marketable for fuel, barbecue purposes, and specialty furniture and flooring. Its pods can be used for baking and to produce alcoholic beverages. Agroforesters are attempting to find methods of converting low-valued, mesquite brush fields to savannah-like stands to produce a variety of products and other benefits to local people.

BENEFITS AND LIMITATIONS

Agroforestry practices are implemented to obtain a higher, more diversified, and more sustainable level of total production than would be possible with a single land-use practice. However, both potential benefits and limitations of a proposed agroforestry practice must be considered. It is important to assess the advantages of an agroforestry practice over monoculture cropping methods and to ascertain the appropriateness of implementing agroforestry at a particular locale.

Benefits

Agroforestry practices provide many goods and services to people. Depending upon the situation, these practices can:

- Use natural resources more efficiently. For example, several vegetative layers utilize solar radiation inputs more efficiently than a single layer; different depths of rooting systems increase cycling of nutrients; and when livestock are included, unused primary production (forage plants) is utilized for secondary production (livestock).
- Increase the level of total crop yields and improve the quality of food production through a diversification of crops.
- Produce primary wood materials from trees and shrubs for the landowner's subsistence, sale in local markets, or export.
- Protect and improve the productive potentials of soil resources.
- Improve socioeconomic conditions by creating employment opportunities, increasing incomes, and reducing the risks of food production and monetary losses.
- Provide a diversity of land-use practices that combine modern technology and local experience and are compatible with sustaining the cultural and social life of the people involved.

Limitations

Limitations to the implementation of agroforestry practices must be considered and overcome. These limitations include:

- Competition of trees and/or shrubs with food crops for sunlight, soil moisture, and nutrients, which can reduce food crop yields.

- Rapid regeneration of prolific tree and shrub species, which can displace food crops and forage plants and take over entire agricultural fields and pastures.
- A potential for trees and shrubs to become hosts to insects and diseases, which are harmful to food crops and forage species.
- Adverse effects to plant growth through the introduction of chemical substances from one agroforestry component to another, a process known as *allelopathy*.
- A requirement of more labor, resulting in a scarcity of a labor force for other agricultural activities.
- Resistance by landowners to displace needed food crops with trees and shrubs, especially when land is scarce.

It is not easy to analyze the benefits and limitations and, therefore, the advantages of a proposed agroforestry practice. The most straightforward method is comparing the benefits and limitations (costs) of the agroforestry practice with traditional livestock production, forestry, or agriculture monoculture crop production (Box 5.2). A problem encountered in making such a comparison is expressing multiple resource values, including ecological benefits and costs, in monetary terms. While some resources can be valued in a marketplace, others can only be valued in social terms. Social values are mostly subjective in nature and vary with the perceptions of concerned stakeholders; therefore, they are difficult to quantify.

Box 5.2

Land Equivalent Ratio: An Approach to Assessing Advantages of Agroforestry (Mead and Willey 1980)

One approach to assessing the advantages of agroforestry is through the land equivalent ratio (LER). LER is a measure of the relative land area that would be required for a traditional monocultural production to attain the same level of production that results from agroforestry, when each operates at the same levels of management inputs (Newman 1990). LER is a ratio between the sum of the land area allotted to each of the components in an agroforestry production and the total land area allotted for a monoculture production. That is:

$$LER = \sum_{i=1}^{n} \frac{Y_{mi}}{Y_{pi}}$$

where Y_{mi} = yield of crop component in agroforestry
Y_{pi} = yield of same crop component in monoculture
n = number of crop components in agroforestry
i = crop component

There is a yield advantage in agroforestry when LER is greater than one. However, the LER is not a measure of economic value. Economic value is measured using the income equivalent ratio (IER). The IER is the ratio of the area needed for a monocultural cropping to the area that is needed for agroforestry practices to produce the same level of income at the same level (intensity) of management. However, it should be recognized that the outcome of any benefit-cost comparison of monoculture cropping and agroforestry is largely dependent on the relative prices of the component crops.

SUMMARY

Agroforestry is a collective term for land-use practices that maintain or, in many situations, increase total yields by combining crops from trees, shrubs, and/or vines, food crops, and livestock production on the same piece of land (either simultaneously or alternatively). It applies management practices that are suitable to the ecological and economic conditions of the area. A variety of agroforestry practices can be implemented to provide stable and sustainable benefits to the landowners.

After reading this chapter, you should have a better understanding of agroforestry and be able to:

- Appreciate the importance of agroforestry as a land-use practice and discuss its importance in relation to problems often encountered in monocultural cropping systems.
- Explain how agroforestry practices are classified in terms of their structural, functional, ecological, and socioeconomic characteristics.

- Discuss the agroforestry functions of windbreaks, silvipastoralism, alley cropping, riparian buffer strips, forest farming, and home gardens.
- Describe the benefits and limitations of agroforestry in general terms.

REFERENCES

Budd, W. W., I. Duchhart, L. H. Hardesty, and F. Steiner, editors. 1990. Planning for agroforestry. Elsevier, New York.

Ffolliott, P. F., K. N. Brooks, H. M. Gregersen, and A. L. Lundgren. 1995. Dryland forestry: Planning and management. John Wiley & Sons, New York.

Forbes, R. G., editor. 1961. Forestry handbook. The Ronald Press, New York.

Gordon, A. M., and S. M. Newman, editors. 1997. Temperate agroforestry systems. Oxford University Press, New York.

Jarvis, P. G., editor. 1991. Agroforestry: Principles and practices. Elsevier Science Publishers, Amsterdam, The Netherlands.

Kidd, C. V., and D. Pimentel. 1992. Integrated resource management: Agroforestry for development. Academic Press, San Diego, California.

MacDicken, K. G., and N. T. Vergara, editors. 1990. Agroforestry: Classification and management. John Wiley & Sons, Inc., New York.

Mead, R., and R. W. Willey. 1980. The concept of a land equivalent ratio and advantages from intercropping. Experimental Agriculture 16:217–228.

Nair, P. K. R. 1993. An introduction to agroforestry. Kluwer Academic Publishers, Dordrecht, The Netherlands.

Rietveld, W. J., technical coordinator. 1995. Agroforestry and sustainable systems: Symposium proceedings. USDA Forest Service, General Technical Report RM-GTR-261.

Rietveld, W. J. 1996. Agroforestry in the United States. USDA Forest Service and USDA Natural Resources Conservation Service, Agroforestry Notes AF 1.

Vergara, N. T. 1981. Integral agroforestry: A potential strategy for stabilizing shifting cultivation and sustaining productivity of the natural environment. Environment and Policy Institute, East-West Center, Honolulu, Hawaii.

Wojtkowski, P. A. 1988. The theory and practices of agroforestry design. Science Publishers, Inc., Enfield, New Hampshire.

6

Wildlife Management Practices

WILDLIFE MANAGEMENT includes the management of wildlife populations and their habitats and the management of people in relation to wildlife in a particular area. The management of people is often a more difficult phase of wildlife management. It can include educational efforts, politics, and the enforcement of regulations on the use or protection of wildlife species. Regardless of its primary focus, wildlife management must be coordinated with the management of other natural resources and land management.

DEFINITIONS

Wildlife are terrestrial animals that are neither human nor domesticated. *Wildlife management* is the art and science of obtaining and sustaining desired populations of wildlife species; it is a form of applied ecology. *Ecology* is the study of the interrelationship of organisms with their environment. Physical and biotic forces affecting an organism or interacting in an area is an organism's *environment*. The place where an organism lives is its *habitat*. An assemblage of populations living in an area is a *biotic community*. The *ecological niche* is the status or role of an organism in the biotic community.

ECOLOGICAL BACKGROUND FOR WILDLIFE MANAGEMENT

Knowledge of terrestrial environments and wildlife population dynamics provides an ecological background that is necessary for the effective management of wildlife and their habitats. Wildlife habitats are found in terrestrial environments. How well wildlife are able to live and survive in their habitats is reflected by how their population changes (fluctuates) in response to habitat conditions. Thus, consideration of terrestrial environments and population dynamics, although only briefly discussed here, is warranted before beginning a discussion of wildlife management practices.

Terrestrial Environments

Vegetation, water, soil, and space are components of terrestrial environments. Arrangements of these components comprise the habitats for wildlife populations relative to the availability of food and cover and to a species' mobility and home range.

Food Supplies

Many wildlife species eat a wide variety of foods. But, a relatively low number of food items have a high nutritional value in any one habitat. A wildlife species may select a particular food because of its availability and not because of its nutritional value. Therefore, it is necessary to determine the frequency of food selection in terms of both availability and the nutritional value. Knowledge of preferred foods of wildlife is a requirement for effective wildlife management.

Mobility

Mobility, or movement, of a wildlife species within a habitat or from one habitat to another varies with the condition of the habitat. If a habitat condition is poor, the need for a species to move, which can increase its survivorship, is high, and vice versa. Mobility of a wildlife species can also vary daily, seasonally, and annually. In general, the larger the size of the wildlife species, the greater is its mobility. Knowledge of the mobility of a wildlife species helps to determine the total size of the area requiring managerial input; the greater the mobility of a wildlife species, the larger the area of management.

Home Range

Each wildlife species requires a specific combination of vegetation, water, and space within its *home range.* The size of a wildlife species' home range is determined largely by its mobility; in general, the greater the mobility, the larger the home range. Interspersion of habitat components, that is, the availability of food, cover, and water on the home range, is crucial to the well-being of a wildlife species.

Habitat boundaries are critical to population management because it is here that the greatest diversity of microhabitats and wildlife species are found. Home range boundaries for a wildlife species change with succession patterns and land-use practices. One objective of wildlife management can be to increase the boundaries of wildlife habitats through practices that alter succession and land use.

Population Dynamics

Population dynamics are fluctuations in the number of individuals in a population and the factors that control these fluctuations. Factors that are key concerns in wildlife management are properties specific to a species, population, or that cause changes in population density.

Species Properties

Reproductive and survival potential are two of the most important species properties. The *biotic potential* of a wildlife species is its reproductive potential plus its survival potential minus the decimating and welfare factors acting on the species. *Decimating factors* kill a wildlife species directly by predation, disease, hunting, or accidents. *Welfare factors* kill a wildlife species indirectly, by reducing the breeding rate and the number of young in a litter. A decrease in the quality of a habitat generally increases the impact of welfare factors.

Limiting factors are those factors that most constrain the well-being of a wildlife species, including the availability of food, cover, or suitable breeding sites. In regions with seasonal changes in weather patterns, limiting factors can change on a seasonal basis.

Population Properties

A wildlife manager should know the age distribution, sex ratio, and birth rate of the population being managed. An *age distribution* indicates whether a population is stable (normal), declining, or expanding at a specific point-in-time (Figure 6.1). Knowledge of the minimum and maximum breeding age helps in the interpretation of an age distribution. Information about the *sex ratio* of a population, when combined with information on mating habits and minimum and maximum breeding ages, provides insight into the growth patterns of the population. Whether the wildlife is a polygamous or monogamous species should also be known.

Birth rates are used to determine the reproductive potential of a population. Input of young into a population is dependent on the number of young in a litter and the number of litters per year. *Mortality rates* are controlled by the survival potential of a population and habitat condition. High mortality rates often are a reflection of poor habitat conditions.

Density Relationships

An S-shaped (sigmoidal) curve is often used to illustrate changes in population density over time. Density is relatively low at the lower portion of the curve, increases in the middle portion, and levels off and occasionally declines in the upper portion (Figure 6.2). The observed changes in population density result from

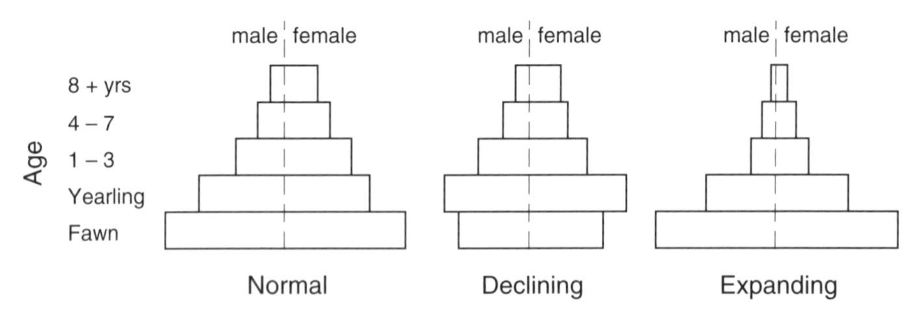

Figure 6.1. Age pyramids indicating stable, declining, and expanding mule deer populations in Arizona (from Hungerford et al. 1981).

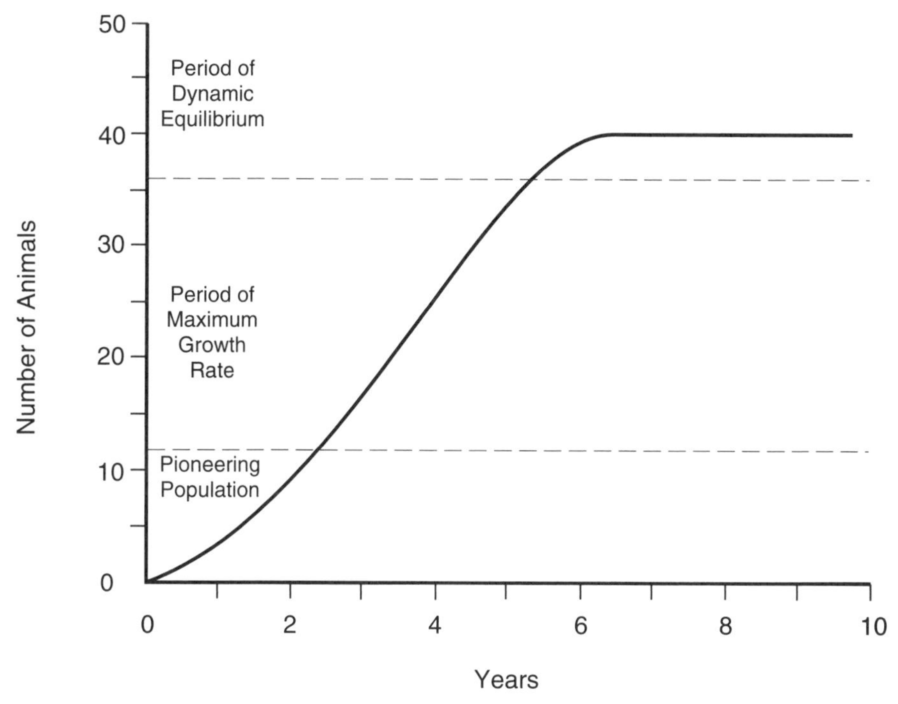

Figure 6.2. A growth model for a hypothetical wildlife population (adapted from Patton 1992).

the simultaneous occurrence of several factors, including species interactions, genetic traits (species characteristics), disease, predators, hunting pressure, and changes in habitat conditions. After these factors become fully effective, a point is reached where the wildlife population levels off and approaches a point of equilibrium with its environment. Once this equilibrium is attained, the upper part of the curve can assume several shapes in addition to the one shown.

The *security threshold* of a population is the minimum number of animals that an area can support without disrupting the biotic potential of the species to a point where the population cannot be sustained. The *saturation point* is reached when an area consistently fails to support more than a certain number of animals. The *range capacity* is the maximum number of animals that can survive on an area for a specified period of time without causing a downward trend in forage production and quality.

POPULATION MANAGEMENT PRACTICES

Wildlife management includes practices that focus on *population management* and *habitat management*. These two phases must be linked together in an integrative manner to manage wildlife effectively and efficiently. Depending on the wildlife species and situation, population management practices involve:

- Specifying hunting regulations to balance a wildlife population with the consumptive value, that is, balance a game species population with its habitat condition.
- Initiating predator control to remove a decimating factor (the predator) that is impacting a (prey) wildlife population.
- Establishing land reservations (refuges) to reduce the impacts of a decimating factor (hunting) or preserve critical habitat conditions.
- Artificially replenishing a wildlife population, when necessary, to sustain the population.

Hunting Regulations

Hunting can be important in the management of a wildlife population. It represents a management tool that can be applied quickly to adjust the size of the population when density has reached the saturation point. However, it is critical that the appropriate hunting regulations be specified to meet this managerial need.

Biological Considerations

The natural turnover (births and deaths) of a wildlife population must be considered in relation to the security threshold when formulating hunting regulations for a wildlife population. The security threshold could be high in the summer and autumn, but bottlenecks (constrictions in biological potentials) can occur in the winter and early spring and reduce the security threshold. Therefore, during these seasons hunting is one method of utilizing a wildlife resource that would otherwise be lost to predation or disease.

The population birth rate and sex ratio of polygamous wildlife species (for example, white-tailed deer) are important factors that determine the most effective hunting regulations. Only killing males may not significantly affect a population's growth because one healthy male can service a number of females. Therefore, in some cases hunting of females becomes necessary to keep a population in balance with the range capacity.

The range capacity of a wildlife population is exceeded in many situations regardless of the wildlife management effort. Controlled hunting can be used to remove the excess animals in these situations. However, a careful examination of the habitat must be made before the hunting regulations are finalized.

Hunting regulations can be formulated to keep a population on the lower, steeper portion of the population's growth curve. This will result in a maximum reproductive gain in the following season or year. Determining proper hunting regulations should also include consideration that the hunting of one wildlife species has implications for other species in the ecosystem. To illustrate this point, one part of a pinyon-juniper woodland in the western United States can be favorable habitat for mule deer, while another part of the woodland is suitable for peccary. Both habitats must be managed properly to achieve proper ecosystem management, which can require a different set of hunting regulations for the two wildlife populations.

Applications

Hunting regulations should be formulated to provide equal opportunities for all hunters. Unfortunately, it is becoming increasingly difficult to attain this managerial goal. Many hunting areas have a saturation point for the number of hunters that can be supported. The sportsmanship of hunters begins to breakdown when this point is reached. Possible solutions to this problem include:

- Adoption of a "first come, first served" policy for issuing hunting licenses and permits.
- Lotteries for the allocation of hunting licenses and permits.
- Purchase of lands for controlled hunting.

A need to control the "kill" of a wildlife population is an essential part of defining hunting regulations. Reducing or increasing the number of animals to be removed from a wildlife population through hunting can be accomplished by:

- Shortening or lengthening the length of a hunting season.
- Reducing or increasing the targeted number of animals to be killed.

- Closing lands on which hunting is permitted or opening additional lands to hunting to increase the number of animals to be killed.
- Specifying that only the males of a polygamous wildlife species or that any sex can be hunted.
- Applying a combinations of these control measures when establishing hunting regulations for a wildlife population.

Some managers feel that most wildlife populations cannot be overhunted, because people will quit hunting when the number of animals becomes so low that hunting is too difficult and too time consuming. The application of proper hunting regulations is intended to ensure that this situation will not occur.

Predator Control

Predator populations have often been controlled. In the past, predators were controlled because hunters felt cheated by the loss of a wildlife prey species or because livestock were being taken. Some predator species were managed because of their economic value (for example, fur). Regardless of the reason, predator control is a complex managerial issue involving various biological problems and applications.

Biological Considerations

A question that should be asked and answered when considering a predator control program is how the population size of prey species will change. For example, the natural control of the prey species can be released with the removal of the predator species (Box 6.1). Predator control programs often push the predator to the steeper portion of the population's growth curve, only to have the population achieve a maximum reproductive gain in the following year. The same principles of population dynamics that apply to prey species also apply to a predator population.

Box 6.1

Predator-Prey Relationships on the Kaibab Plateau in Northern Arizona (Rasmussen 1941, Patton 1992)

Before the removal of predators by a control program was initiated in 1906, the Kaibab Plateau had an estimated

4,000 mule deer (the prey). Mountain lions, wolves, coyotes, and bobcats (the predators) were then removed by paid hunters and trappers between 1906 and 1923. Twenty years later, before a major die-off due to the imbalance between predator and prey numbers, the mule deer population was estimated to be 100,000. Wildlife biologists concluded that due to the control of the predators the mule deer population had expanded well beyond the range capacity of the Kaibab Plateau, even though 200,000 sheep and 20,000 cattle were removed from the plateau in the same period. Several decades passed before the mule deer population once again was balanced with its habitat.

The natural turnover of the prey population is important in evaluating the effectiveness of predator control. For instance, predators can mainly remove the fraction of the prey population that would be lost through natural mortality. Therefore, it is necessary to consider the reproductive potential of both the prey and predator populations in evaluating the effectiveness of predator control. In general:

- When the reproductive potential of a prey population is less than that of the predator population, control of predator pressure can cause the prey to become more abundant, although this can be the objective of a predator control program.
- When the reproductive potential of a prey population is greater than that of a predator population, predator control is likely to have little impact on the prey population.

Food habits of the predator need to be known before initiating a predator control program. Whether the predator is actually eating large quantities of the prey must be determined. For example, coyotes have been controlled in some instances to release mule deer populations, although coyotes mainly eat small rodents and rabbits.

At times, the so-called "sanitation theory" has been used to justify retaining rather than controlling predator populations. This theory states that diseased and deformed prey are the animals most commonly removed by predators and only the healthiest of the prey survive. It is questionable that the theory of sanitation is valid.

Most predation appears to be a result of opportunity and circumstance.

Applications

Historically bounties have been used to remove selected predator species. Some bounty hunters have abused the use of bounties, taking only the males of a predator population and leaving females to reproduce and, thus, provide a continued animal supply for bounty collection. Removal of predators by bounties rarely pushes the species below the reproductive part of the population growth curve. Paid trappers or trapping by management agency personnel has been more successful in getting a predator population to the lower part of the population density curve. Attempts to poison predators are generally too indiscriminate, with species other than the targeted predators being killed. Also, it is dangerous.

Predator control should be applied only on a case-by-case basis and only when it is known that a predator is reducing the productivity of its prey to the point where economic losses are excessive.

Land Reservations

Land reservations are often established to reduce the impacts of hunting on a wildlife population (hunting is generally not allowed on land reservations). It is assumed that these areas will provide an excess of wildlife that will "flow out" from the land reservations to inhabit the surrounding areas.

Biological Considerations

Many people like the idea of land reservations because their establishment is thought to represent a form of conservation. To other people, land reservations offer special privileges to those who hunt around the edges of the reservations. Biological problems can also occur; therefore, they must be considered in the establishment of land reservations. Home ranges of certain wildlife populations must be larger than the land reservations to ensure that the species flow out into surrounding areas. A wildlife population that is territorial in its habitat use can also cause the species to flow out of a land reservation. The tolerance of a species to increases in its population density and the higher population density of other wildlife species must also be considered.

Habitat conditions on land reservations can change over time, primarily through plant succession. Therefore, appropriate habitat management must be practiced to achieve desired results. Land reservations should be integrated into the landscape of surrounding areas. A forested reservation in the middle of a large grassland ecosystem can be unacceptable if the wildlife population of concern requires a forest and its home range is larger than the reservation.

Applications

Land reservations are often established to protect a threatened or endangered wildlife species. Establishment of land reservations can also be a key to the management of migratory waterfowl. But, land reservations could "lock up" a wildlife population that does not require protection. In other instances, overhunting can occur around the edges of land reservations, defeating the purposes of their establishment. Crop damage occurs when wildlife protected by land reservations flow out and detrimentally impact surrounding agricultural areas. Range capacity problems are created when the home range of a wildlife population is larger than the land reservation, making it impossible for the species to flow out into the surrounding areas.

Land reservations are also established for habitat improvement or development. "Habitat improvement" reservations have become more common than "no hunting" reservations in many regions of the United States. Land reservations should be controlled by one wildlife management agency or organization and managed in one block, if possible. In some instances, land reservations have been converted into "public shooting grounds" but generally with limited success.

When land reservations are established, the boundaries of the reservations must be clearly marked. Trespassing by people onto reservations also needs to be controlled.

Artificial Replenishment

Artificial replenishment involves the introduction of a wildlife species into an area where the species had not been found previously or the reintroduction of a wildlife species into an area where the species originally occurred. Artificial replenishment can also include maintaining a species at a specific population size that is consistent with habitat conditions.

Introduction

Introduction of a wildlife species has met with mixed success and, therefore, must be considered with caution. Interrelationships between the introduced wildlife species and indigenous wildlife species and the impact of the introduction on existing habitat conditions must be considered. Prior to introduction, it should be known whether species-specific diseases and parasites are also being introduced. In some cases, introductions of wildlife species had detrimental impacts on sustainable livestock and agricultural crop production. Therefore, the introduction of a wildlife species should not be adopted as a management practice without first evaluating small-scale introductions.

Introduction of an indigenous wildlife species to augment a depleted population can be justified in situations where habitat conditions have been severely disturbed by improper natural resources management practices. However, introductions for this or other purposes must be monitored carefully to minimize potential conflicts in habitat selection with the indigenous wildlife species. Such conflicts are pronounced when indigenous species are attempting to reestablish themselves after depletion.

Reintroduction

Reintroduction of a wildlife species differs from wildlife introductions, although it has also been a cause of controversy when local people object to the practice (Box 6.2). Successful reintroduction of a wildlife species requires establishing appropriate release areas based on the habitat requirements for the wildlife species involved. Holding pens and breeding paddocks must also be constructed, if needed. Proper release procedure requires selecting the best time, space, and appropriate number of animals to be released at one time. Local people should be informed about the release area locations and the number of animals that will be released. Before being released, animals should be marked. This will allow for easy identification when monitoring efforts are made to evaluate the success of the reintroduction.

Maintenance

A wildlife population that has declined to a point where it is approaching its security threshold can be increased through artificial replenishment. Artificial replenishment can be used to maintain the population at a level that is consistent with the habitat conditions. Before maintenance is initiated, any physiological differences between the animals to be released and the resident wildlife population should be assessed. Physiological differences could lead to low survival rates for the released animals.

Box 6.2

The Mexican Wolf: A Reintroduction Controversy (Parsons 1998)

Predation on livestock by the Mexican wolf (*Canis lupus baileyi*) increased greatly in the late 1800s and early 1900s. Predation by wolves increased as the number of livestock in the southwestern United States increased and native ungulates, the primary prey of the wolf, declined due to unregulated subsistence and market hunting. Pressure, largely from ranchers, to reduce and in some cases to eliminate the wolves resulted in aggressive trapping, shooting, and poisoning. By the middle of the 1900s, the animal was extirpated from the region. It was not surprising, therefore, that in 1976 the Mexican wolf was listed as endangered under provisions of the Endangered Species Act (see below). In response to this listing, the U.S. Fish and Wildlife Service formed a *Mexican Wolf Recovery Team* in 1979. The purpose of the team was to ensure the survival of the animal by maintaining a captive-breeding program and to reestablish a viable, self-sustaining population (100 wolves or more) within the Mexican wolf's historic range.

Eleven wolves from three families were "soft released" after about two months of acclimation onto the Apache National Forest in eastern Arizona in March 1998. A soft release involves holding the wolves in pens on-site prior to release. This allows them to become familiar with the release area and reduces their tendencies to disperse following release (Fritts et al. 1997). Whether this and subsequent releases in 1999 will ever be successful and, if successful, accepted by local people in the long run is debatable. Ranchers in the area fear that their livestock will be lost to Mexican wolf predation. In attempting to sustain the reintroduction effort, *Defenders of Wildlife,* an environmental advocacy organization, have compensated livestock owners for known incidents of livestock depredation. Nevertheless, the popular press reported that some of the wolves in the initial release were killed by mysterious circumstances (Murr 1999). Because of the continuing debate over the worthiness of the program and in response to complaints from ranchers following attacks on livestock, authorities in charge of the Mexican wolf reintroductions began capturing and relocating the released wolves in the summer of 1999 to habitats in neighboring New Mexico. Whether the wolf will ever be returned to the wilds of the southwestern United States is unknown.

Restocking of a wildlife population for maintenance can be implemented on a put-and-take basis, where the number of animals that are released equals the number of animals lost. This practice is often costly, however. Nevertheless, the put-and-take method of restocking can be justified in situations where the hunting pressure is high and hunters agree to pay for the practice.

HABITAT MANAGEMENT PRACTICES

Wildlife populations require food, protective cover, water, and opportunities to reproduce and nurture their young. Therefore, a principal stage of wildlife management is habitat management. In many cases, habitat components are altered by natural resources management practices more readily than are actual populations of wildlife species.

Properly planning the planting of trees, shrubs, or herbaceous plants; increasing the production of shrubs and herbaceous plants by reducing the density of competing forest overstories through timber harvesting; or converting woody communities to grasslands or other herbaceous covers enhances habitat conditions for some wildlife populations. However, these same vegetation management practices can adversely impact the habitats of other populations.

Principles

Some wildlife populations in particular habitats have a considerable impact on the other wildlife populations present. When this is the case, initial management concerns can be for habitat improvement that benefits the most dominant populations. One species will often use more than one habitat type. The home ranges of large ungulates (e.g., deer, elk, or antelope) can include grazing in a number of habitats, while migratory animals frequently move across various habitat types (and political boundaries) in their life cycle. When a wildlife manager confronts such instances, the complexity of all habitat types has to be considered jointly, which makes management difficult.

The habitat requirements throughout the lifetime of a wildlife species must be known. Species with long life spans often have a greater effect on their environment than species that live only for one or a few years. Older animals can have different habitat requirements than younger animals. Males of a species can also differ from females in their habitat requirements. Therefore, knowledge of habitat requirements for wildlife populations of differing ages and sex is a key to sustainable wildlife management.

Removal of excess wildlife from an area becomes necessary when the population is larger than the carrying capacity of the habitat. Excess wildlife can be live-trapped or taken with drugs and then moved to areas where populations of that wildlife species are relatively low. Wildlife populations can also be harvested through controlled hunting. When the latter is practiced, it is necessary to specify which species and the numbers of each species that should be removed by hunting in order to regain a balance between the wildlife and their habitats.

Applications

Habitat management is largely based on knowledge of plant succession of a managed area (Figure 6.3) and the managerial tools

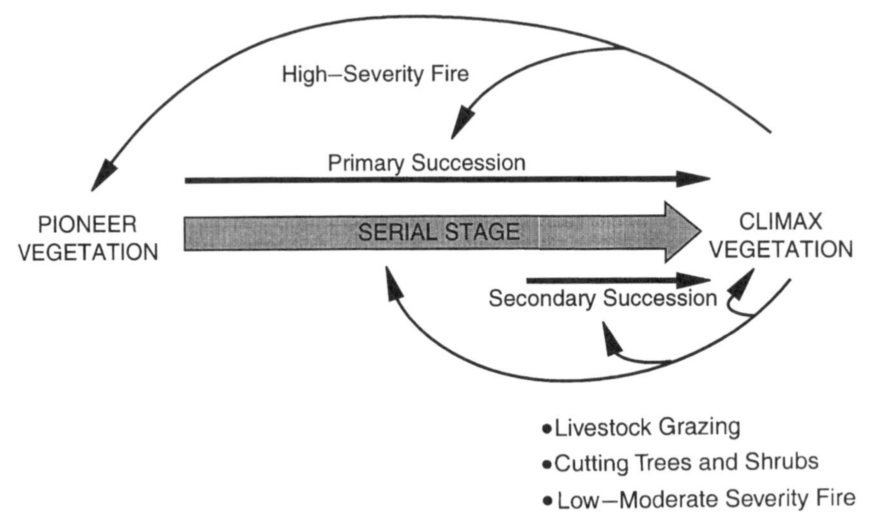

Figure 6.3. Community replacement progresses to a climax plant community in the classical Clementsian concept of succession. *Primary succession* is initiated on lava flows, sand dunes, and other newly exposed sites. *Secondary succession* follows disturbances such as fire, excessive livestock grazing, or cutting of trees and shrubs (adapted from DeBano et al. 1998).

that affect plant succession. Livestock grazing, cutting of trees and shrubs, and prescribed fire can be used to slow down plant succession to obtain a desired habitat condition. When it is desired to speed up the rate of plant succession, a wildlife manager can protect an area from livestock grazing and fire in combination with the planting of trees, shrubs, or herbaceous plants. Variation of habitat conditions is often the purpose of habitat management, requiring both the speeding up and slowing down of plant succession.

In the past, many natural resources management practices tended to create a monotype of habitat conditions. However, the goal of more recent management practices is often to achieve a greater diversity of habitats. It is often assumed that the greater the biological diversity, the greater the diversity of wildlife species.

MANAGEMENT OF ENDANGERED WILDLIFE SPECIES

One perspective on conserving wildlife species in the United States is contained in the *Endangered Species Act* passed by the

Congress in 1973. This act provides for the conservation and recovery of wildlife species designated as endangered. A species is *endangered* if it is about to become extinct throughout a significant part or all of its range. A *threatened* wildlife species is one that is likely to becomes endangered in the foreseeable future. Designation of an endangered or threatened status for a wildlife species is determined by the federal government, specifically by the Secretary of the Interior. Similar forms of legislation have also been introduced by states and in other countries.

Some ecologists have estimated that one wildlife species in the world becomes extinct each year, while others of a more pessimistic view believe that if plants and invertebrates were included in this category, future extinction rates would increase to one per day. Choices about which wildlife species to save or perpetuate through management are becoming increasingly difficult. It is often easier to generate concern about a wildlife species, such as the Sonoran pronghorn, that people can see in the wild than it is to generate concern for a species as small and "nonfurry" as the Texas salamander. In either case, a wildlife manager must be responsive to the public and must implement the necessary management practices that meet the expressed concerns.

SUMMARY

Wildlife management includes practices to obtain or maintain populations at desired levels and practices that ensure provision of adequate food, cover, and water for wildlife. A wildlife manager must link these two phases of management to manage wildlife effectively and efficiently. In essence, wildlife management is a form of applied ecology.

When you have finished reading this chapter, you should understand:

- The importance of species properties, environmental properties, and limiting factors in studying population dynamics.
- The relationship between the home range of a wildlife species and the population management of that species.
- The manner by which hunting regulations, predator control, land reservations, and artificial replenishment are applied in wildlife management.

- The principles of habitat management and the way by which plant succession can be modified to obtain desired habitat conditions.
- The importance of managing populations and habitats of endangered and threatened wildlife species.

REFERENCES

Cappuccino, N., and P. W. Price. 1995. Population dynamics: New approaches and synthesis. Academic Press, Inc., Orlando, Florida.

Caughley, G., and A. R. E. Sinclair. 1994. Wildlife ecology and management. Blackwell Scientific Publications, Boston.

DeBano, L. F., D. G. Neary, and P. F. Ffolliott. 1998. Fire effects on ecosystems. John Wiley & Sons, Inc., New York.

Fritts, S. H., E. E. Bangs, J. A. Fontaine, M. R. Johnson, M. K. Phillips, E. D. Koch, and J. R. Gunson. 1997. Planning and implementing a reintroduction of wolves to Yellowstone National Park and central Idaho. Restoration Ecology 5:7—27.

Hungerford, C. R., M. D. Burke, and P. F. Ffolliott. 1981. Biology and population dynamics of mule deer in southwestern United States. In: Ffolliott, P. F., and S. Gallina, editors. Deer biology, habitat requirements, and management in western North America. Instituto de Ecological, Mexico, D. F., pp. 109–132.

Leopold, A. L. 1986. Game management. University of Wisconsin Press, Madison, Wisconsin.

Morrison, M. L., B. G. Marcot, and R. W. Mannan. 1997. Wildlife-habitat relationships: Concepts and applications. University of Wisconsin Press, Madison, Wisconsin.

Murr, A. 1999. Deadly days for wolves: An animal murder mystery unfolds in the West. Newsweek (November 30, 1998), p. 34.

Parsons, D. R. 1998. "Green fire" returns to the Southwest: Reintroduction of the Mexican wolf. Wildlife Society Bulletin 26:799–807.

Patton, D. R. 1992. Wildlife habitat relationships in forested ecosystems. Timber Press, Portland, Oregon.

Peek, J. M. 1986. A review of wildlife management. Prentice Hall, Englewood Cliff, New Jersey.

Rasmussen, D. I. 1941. Biotic communities of the Kaibab Plateau. Ecological Monographs 3:229–275.

Schemnitz, S. D., editor. 1980. Wildlife management techniques. The Wildlife Society, Washington, D.C.

7

Fishery Management Practices[1]

FISHERY MANAGEMENT involves the use of scientific information to decide how to manipulate aquatic environments, aquatic biota, and human activities so that targeted populations of organisms and economic and environmental conditions are sustained. Fishery management is similar to terrestrial wildlife management (see Chapter 6) because populations of fishery species are a renewable natural resource and are generally considered a public resource. Fishery management involves the manipulation of biota, environmental conditions, and human behavior, while recognizing that the maintenance of viable populations is dependent upon wise management of land and water. An added element to fishery management has been the growing concern for the conservation of native species, including their genetic diversity. However, fishery management differs from wildlife management because management sometimes is linked to a major food industry (e.g., commercial harvest of fish and aquacultural production) and the assessment of populations largely depends upon sampling unseen animals.

Management of self-sustaining populations of fish and other biota (including shellfish and other invertebrates, algae, vascular plants, amphibians and reptiles) occurs in both marine and freshwater environments. The focus of this chapter is management of

[1]William J. Matter, University of Arizona, Tucson, is a joint author of this chapter.

freshwater fish. However, many of the concepts presented also apply to the management of other freshwater and marine biota.

DEFINITIONS

Fish are classified as those harvested commercially, those harvested for sport (*game species*), and those not harvested (*nongame species*). Historically, a *fishery* has been defined as a system made up of a fish species of interest to humans, the aquatic environment that supports the species, and the harvest of a targeted fish species (e.g., the recreational walleye fishery in Lake of the Woods, Minnesota). However, today fishery managers are equally concerned about species that are not harvested. *Aquaculture,* the husbandry of captive aquatic organisms, can be thought of as a component of fishery management, but more often aquaculture is considered as a special type of agriculture. Effective *fishery management practices* require knowledge of aquatic ecosystems and the target organisms. Important related subjects include ichthyology, limnology, aquatic ecology, fish biology, and conservation biology. Fishery management involves interactions among aquatic organisms, environmental conditions, and human activities, but for convenience this chapter will focus on each area separately.

ECOLOGICAL BACKGROUND FOR FISHERY MANAGEMENT

Knowledge of freshwater environments and fish population dynamics is important and necessary background information for the effective management of fish. Freshwater environments provide fish habitat, while the ability of fish to thrive in their habitats is reflected by how their population changes (fluctuates) in response to changes in habitat conditions. Thus, a discussion of freshwater environments and population dynamics is warranted before considering fishery management practices.

Freshwater Environments

Management of freshwater fish is practiced in natural lakes, ponds, streams, rivers, as well as in human-made reservoirs built for flood control, recreation, hydropower generation and in structures that store water for irrigation, industry, and human consumption and use. The characteristics of freshwater environments of most interest to fishery managers are water quality, especially temperature, oxygen content, and nutrient content, and the availability of required habitat features such as cover, substrate, and food supply for fish and other aquatic animals.

Temperature

Temperature largely determines the fish species that can thrive in a water body. The range of temperatures that a fish species can tolerate is much broader than the range of temperatures preferred. Selected fish species are grouped by their preferred water temperatures in Table 7.1. The existence of certain species is threatened when temperature tolerances are exceeded or when temperature rapidly changes. Temperature can increase when riparian vegetation is removed by livestock grazing or harvesting of trees or when heated wastewater is discharged into a waterway. For a stream located below a reservoir or dam, temperature declines when cold water is released from the lower depths of the reservoir. For example, the Colorado River in much of the desert southwestern United States is now much colder in summer and warmer in winter than occurred historically because deep, cold water is released from the reservoirs along the river. Cold water released into areas below dams has contributed to the decline of native Colorado River fishes and to the persistence of non-native salmonids. Fish will move away from water bodies that have temperatures unfavorable for growth and survival.

Fertility

Fertility is a measure of the presence of nutrients needed to sustain primary productivity (Table 7.2). Water that is infertile is unlikely to sustain fast growth and high densities of fish, while

Table 7.1. Preferred temperatures of selected fish species

Category	Preferred Temperatures	Example Species
Cold-water Species	10–15°C (50–60°F)	trout, salmon
Cool-water Species	20–25°C (68–77°F)	walleye, northern pike, striped bass
Warm-water species	25–30°C (77–86°F)	largemouth bass, sunfish, carp

Table 7.2. Levels of total nitrogen (N) and phosphorus (P) in standing water bodies of differing fertilities

Nutrient	Infertile	Medium Fertility	Fertile
N	<0.02 ppm	0.02–0.10 ppm	>0.10 ppm
P	<0.01 ppm	0.02–0.10 ppm	>0.10 ppm

noxious algae blooms, large beds of aquatic weeds, and oxygen depletion are frequently found in water that is too fertile. The occurrence of other elements (silica, iron, potassium, and carbon dioxide) can also be important in determining fertility.

The fertility of lakes and reservoirs is determined largely by changes in available nutrients resulting from:

- The nutrient content of surface runoff or streams flowing into the water body.
- The nutrient content of precipitation falling directly into the water body.
- The rate of recycling of nutrients within lakes and reservoirs during circulation of water (i.e., the turnover).
- The removal of biota (e.g., algae, aquatic vascular plants, insects, or fish), which removes nutrient capital from the water body.

The fertility of streams and rivers is determined primarily by the nutrient content and amount of organic material found in the runoff draining a watershed that enters the stream or river channel. The addition of fine particles that remain suspended in water can block the penetration of sunlight and limit productivity, regardless of the nutrient availability of in a lake or stream.

Dissolved Oxygen

Low dissolved oxygen in water often limits survival or production of fish and can render some aquatic environments uninhabitable. Water in shallow lakes and ponds (less than two to three meters deep), especially if they are infertile, can remain oxygenated by photosynthesis and the surface agitation caused by blowing wind. However, circulation of water and oxygen in deeper lakes can be restricted in summer by stratification of the lake into layers, or *strata,* of different temperature and density. The upper layer of a stratified lake, called the *epilimnion,* is warmer and less dense than deeper, colder layers. The epilimnion can remain oxygenated by the action of atmospheric diffusion, surface wind disturbance, and photosynthesis. Deep water is sealed off from wind and atmospheric exchange. Microbial decay of dead algae, vascular plants, and animals on the bottom of a lake consumes oxygen, and because sunlight does not penetrate to these depths, no oxygen is produced by photosynthesis. The greater the fertility of the lake, the greater is the rate of oxygen depletion. Enrichment of lakes due to human-caused addition of nutrients and organic matter (e.g., sewage discharge or runoff from feedlots and fertilized agricultural land) speeds the depletion of oxygen and can result in fish death and reduction of available habitats for aquatic organisms. In many instances, lakes *turn over,* or mix, in the fall or spring and, as a consequence, the entire lake becomes reoxygenated.

Fish in shallow, fertile lakes in regions with cold winters can undergo *winter kill* because of oxygen depletion under ice cover. Ice limits oxygen uptake from the atmosphere. High demands for oxygen under the ice, especially for microbial decay of organic matter, can lead to oxygen depletion and fish death. Winter kill is seldom a problem in deep lakes of low to moderate fertility.

Oxygen deficiencies are generally less common and less severe in shallow, fast moving rivers and streams because air and water mix continuously. However, low oxygen can occur in relatively deep and slow moving rivers and streams, especially if inputs of organic matter place a large demand on dissolved oxygen.

Food Supply

Food for fish consists of phytoplankton and zooplankton (microscopic plants and animals living in the water column), algae and detritus attached to substrates, benthic invertebrates (e.g., insects), and other fish. Foods selected by fish generally increase in size as fish grow from larvae to adult. Food sources for all life stages of a fish species must be readily available for sustained fish production. Food for each of the prey eaten by a target species of fish (that is, *links* in the *food chain)* must also be available.

Other Factors

Other factors that effect productivity of fish populations in freshwater environments include the condition of spawning areas, availability of refuge habitat for predator avoidance, pollution, and fluctuating water levels.

Each fish species has its own combination of habitat requirements for spawning and rearing of young, growth to maturation, overwintering, as well as other specific life-history stage requirements. These requirements must be met for a species to be self-sustaining. Some species (e.g., bass and sunfish) require clean substrates in which to dig depressions (nests or *redds)* for eggs, while other species (e.g., trout) lay and bury eggs in streambed gravel that has substantial intragravel water flow, and still others (e.g., walleye and northern pike) broadcast eggs in the water column above sediment-free substrates.

Organic pollutants in sewage, industrial effluent, or agricultural or urban runoff can be directly toxic to fish or can indirectly affect fish production by depleting dissolved oxygen levels. Other chemicals (e.g., heavy metals, pesticides, PCBs, petroleum products) are also directly toxic to fish and to fish food organisms. Fine sediments from land areas disturbed by harvesting of timber, livestock grazing, mining, or road construction can accumulate on the bottom of lakes and streams and impair conditions for food production and spawning.

Fluctuating water levels (especially in reservoirs) can limit the productivity of fish by destroying food-producing communities and nesting areas, trapping fish in backwaters, and by causing bank erosion. Rising water levels that flood terrestrial soil and vegetation can increase the fertility of a water body and create new spawning areas for some fish species. Diversion of water out of streams for irrigation, industry, and urban use can reduce the volume of water in a stream to the point where habitable areas are reduced or lost. Knowing the *instream flow* requirements for aquatic organisms to complete their life cycles is often essential for effective fishery management in streams.

Population Dynamics

The principles that govern the population dynamics of terrestrial wildlife populations (see Chapter 6) also apply to fish populations. The size of a fish population fluctuates in response to its reproductive capacity and biotic and abiotic environmental pressures. Limiting factors tend to establish the carrying capacity of a freshwater environment, but competitive interactions between members of the same species and between different species and predation by fish, birds and mammals also influence the abundance of fish.

A fish population typically consists of multiple life stages (larvae, juveniles, subadults, and adults) and multiple age classes of adult fish. The proportion of individuals in each age class varies by species and depends on whether a population is stable, expanding, or declining. Similar to terrestrial wildlife, the size of a fish population also fluctuates through time. Populations tend to grow in numbers until reaching the carrying capacity of the water body, but fluctuations in numbers occur continuously as changes in the environment alter the carrying capacity or cause mortality.

POPULATION MANAGEMENT PRACTICES

Setting regulations for the harvest of fish, control or removal of undesired fish populations, and stocking of fish are among the management practices that are implemented to change the characteristics of a fish population or the composition of a fish community.

Harvest Regulations

In setting harvest regulations, fishery managers must consider fish growth rates, survival, *recruitment,* age of sexual maturity, and the amount of harvest that occurs. Recruitment is the number or percentage of fish that grow large enough each year to be considered catchable or reproductively mature.

For commercial fishing, obtaining the highest sustainable harvest of fish in weight or monetary value is often the purpose of setting harvest regulations. Maintaining sufficient numbers of harvestable-size fish and equally distributing the catch of fish for all anglers are common goals of regulations for sport fishing. Regulating the use of live bait and releasing captive fish can be used to help protect water bodies from introductions of non-native species.

Regulations for harvest of game fish are designed to:

- Reduce angling pressure to avoid overfishing.
- Protect the harvesting of vulnerable life-history stages to avoid overfishing.

- Equally distribute the catch of fish across all anglers.
- Enhance the size distribution of fish within a population.
- Create special angling opportunities, for example, trophy fishing, tournament fishing, fly-fishing only, or wilderness fishing.

Biological Considerations

The formulation of effective harvest regulations requires knowledge of the abundance and age structure of a fish population. To get a complete picture of the population, this information is obtained by sampling with one of a variety of methods including gill netting, seining, electroshocking, trapping, and underwater observation. Ideally, a harvested population should have an adequate number of spawners at the end of the harvesting season to sustain the population. Also, a population should be large enough to provide a suitable catch in relation to the fishing effort expended and have a sufficient number of individuals and the proper age structure to ensure acceptable growth and survival.

Applications

Formulation of effective harvest regulations is difficult largely because recruitment is variable from year to year, due to changing environmental conditions. In general, because larger, older fish are harvested first, the greater the rate of harvest, the smaller and younger are the individuals remaining in the population. Excessive harvest of fish can effectively eliminate, although not necessarily exterminate, a population.

Harvest regulations can designate:

- Size and number of fish to be taken.
- Season of harvest, both timing and duration.
- Type of allowed fishing equipment (or the fishing gear).

Growing numbers of anglers cause higher fishing pressure, which requires more restrictive regulation to prevent overharvest of preferred species. Catch-and-release angling, where all captured fish are released, may be required to sustain angling in water bodies with exceptionally high fishing pressure.

Control of Populations

Control of fish populations is used primarily to reduce abundance of undesirable fish; however, eradication of an undesired species can rarely be achieved. Which fish species are defined as undesirable depends largely on personal viewpoint. Historically, fish with little economic value were frequently considered undesirable.

Biological Considerations

Today fish populations are generally controlled when they are destroying habitat of a valued fish species, competing directly with desired fish for food and cover, or preying on a desired fish species. In some situations, however, populations of undesirable fish represent a potential resource to be managed and exploited. A fishery manager might reduce the numbers of fish in a population in small ponds and lakes or reservoirs when:

- The number of predators and prey is not in balance.
- Large numbers of small, slow-growing (stunted) fish make up the population.
- Habitat conditions are being degraded by undesirable species.
- A non-native species has negatively impacted a native species.

Applications

Destroying spawning areas and nesting sites of undesirable fish is one approach to controlling populations. Predatory fish have also been introduced to control undesirable species, but this method rarely achieves total control and often results in the introduction of a non-native predator, which often has unanticipated negative impacts. Poisoning of all undesirable fish and restocking with desired species is a relatively drastic application that must be thoroughly evaluated before use. It is difficult to completely remove a fish species from a water body. It should also be noted that the accidental or purposeful reintroduction of undesired species by the public frequently nullifies control efforts.

Stocking of Fish

The public often considers the addition of new fish to a water body, that is, stocking, as the best management tool available to fishery managers. However, while stocking has been successful in meeting some objectives of management, it should not be viewed as a panacea but as a tool to be applied carefully in prescribed conditions (Box 7.1). Stocking is often used to start new self-sustaining populations (stocking only once) or to maintain species that are not self-sustaining (repeated stocking).

Box 7.1

Stocking of Fish: Some Considerations

Stocking of fish into water bodies outside of their historic range should be avoided. However, stocking of fish can be appropriate when:

- There is little or no natural reproduction of a managed fish species.
- It is necessary to reestablish a population of fish that has disappeared due to habitat loss or other forms of decimation.
- The rate of local fishing success is greater than the rate of natural reproduction and recruitment sustained by a game fish species.
- There is desire for a local concentration of fish for sport-angling success in a water body, such as a pond in an urban park, near a bridge, or along a road.

After deciding to stock fish into a water body, the fishery manager must determine:

- The proper size of fish to stock, that is, whether fry, fingerling, or catchable-size fish will be stocked.
- The number of fish to stock so that the capacity of the environment to support fish is not exceeded and the anticipated level of angling is met.
- The potential impact of stocked fish on the existing aquatic community.

Biological Considerations

Stocking of fish can be a management option when there is insufficient natural reproduction of a desired species and habitat conditions are suitable for the survival and growth of the species. Stocking is also used if angler demand for a fish species is greater than the population size sustained by natural reproduction and growth. In either case, it is necessary for managers to demonstrate that stocking of fish is the most efficient way to maintain the species in the water body. Stocked fish can compete with other fish species for food and space, or they may prey on other species present in a water body. Stocking of non-native fish is often in conflict with efforts to maintain and enhance native species.

Applications

The proper fish size, time of stocking, and number of fish stocked must be decided upon before stocking is undertaken. Fry, fingerlings, and catchable-size fish can be stocked. Stocking of fry or fingerlings is relatively inexpensive, but these small, young fish usually suffer high rates of mortality after stocking. Catchable-sized fish are often stocked in intensively fished water bodies (e.g., urban lakes and streams). These are referred to as *put-and-take* fisheries because fish are put in and then taken out by anglers. These fish are not expected to undergo growth and reproduction. The time to stock a fish species is often related to the time of year when the water temperature is optimal for the survival of the species (Table 7.1).

The number of game fish to stock depends largely on the anticipated rates of mortality and harvest and the costs of producing fish at a hatchery and delivering them to a site. There is little point in overstocking fish, because excess fish suffer high mortality and impact the survival and quality of other fish. Both economic and ecological consequences of stocking must be considered to determine if anticipated costs are justified.

HABITAT MANAGEMENT PRACTICES

Loss of suitable habitat is the most significant cause of declines in the abundance and occurrence of freshwater fish species. Degradation of habitat conditions is probably the second greatest threat. Habitat management becomes necessary if some aspect of habitat quality prevents potential productivity of a fish population from being realized. However, habitat management can be a waste of time and money if specific conditions limiting productivity (oxygen, nutrients, cover, food organisms) are not known. When a habitat feature is plentiful and not a limiting factor, enhancement of that feature has little value.

Principles

Habitat management practices involve manipulation of the physical and biotic features of the freshwater environment. Examples of these practices include:

- Protecting the surrounding watershed from disturbances that lead to excessive erosion and high rates of sedimentation in lakes and streams.
- Controlling water levels and water quality (especially nutrients and contaminants).
- Protecting or replacing riparian vegetation and controlling bank erosion.
- Adding natural or artificial structures to create shelter and spawning areas in reservoirs, pools, backwaters, riffles, or other hydrologic features in streams.
- Controlling or removing unwanted aquatic vegetation.

Applications

Protecting the surrounding watersheds from erosion to avoid potential sedimentation problems is a topic discussed in Chapters 2 and 11; therefore, they will not be detailed here.

Water levels typically fluctuate in natural lakes and streams, so maintaining habitat for most fish species does not require holding water at a constant level. However, control of water levels in reservoirs can be used to flood shoreline areas to increase spawning areas for a desired fish species, or water levels can be lowered to strand nests of undesirable species. Frequent water level changes can hinder development of aquatic vegetation in the shallow littoral zone. The greatest challenge for managing streams is ensuring that a sufficient volume of water remains in the channel to support the entire life history of fish species.

Increases in water temperature can kill fish and their food and make areas uninhabitable for some aquatic species. Water temperatures can increase after streambank vegetation is cleared because increased solar radiation hits the water surface. Retaining streambank vegetation often helps to maintain appropriate water temperature. Heated effluents from power plants and industries also increase water temperature. Cooling towers or cooling ponds are used to cool water before it is released into a water body, but some heated waste nearly always is carried into receiving waters. Water released from reservoirs can be comparatively cold or warm depending on the depth in the reservoir from which water is released. Some dams have ports at a range of depths so that the temperature of the water released can be adjusted to match natural conditions.

Bank erosion that results from livestock grazing of riparian plants can degrade aquatic environments. Areas adjacent to water bodies can be fenced off to prevent livestock grazing (see Chapter 2). Increasing the density of trees and shrubs surrounding a water body also can act as a barrier to movement of livestock to a water body.

In streams, *pools* are areas of deeper, slow-moving water compared to *riffles,* which have shallow, fast-moving water. It can become necessary to create pools or riffles to enhance conditions for some stream fish. Pools are formed below low dams, and riffles can be created next to artificial deflectors. It is difficult to add structures that are likely to withstand even moderate floods, however. A variety of natural and artificial structures can be added to reservoirs to create *reefs,* where fish can find shelter and spawning substrate. Reefs can replace cover that was once provided by flooded trees and other vegetation but over time decayed or was buried by sediment as the reservoir aged.

Regardless of the habitat management practices imposed, fishery managers should make provisions for periodic assessment of the effectiveness of habitat manipulations and for regular maintenance or replacement of habitat structures. Otherwise, the investment made in enhancing the habitat conditions can be lost.

SUMMARY

Fishery management can be difficult because of the complexity of aquatic ecosystems, and because the many demands from people can impact water availability and quality. Also, knowledge of the properties of fish populations required for management is largely dependent on sampling unseen animals. Therefore, the challenge for fishery managers is to meet the growing demands of people for sustainable fish populations while protecting or enhancing the quality of the aquatic ecosystems that support freshwater and marine biota.

At the end of this chapter, you should be able to:

- Discuss the characteristics of aquatic environments of primary interest to fishery managers.
- Explain the basic principles that govern the dynamics of fish populations and how the multiple life stages of fish species influence population dynamics.
- Describe the problems that fishery managers face when formulating harvesting regulations and explain why recruitment and the age of vulnerability are important in setting harvesting regulations.
- Describe the methods that can be used to control fish populations and the conditions present when it becomes necessary to use each method.
- Discuss the use of stocking as a fishery management practice and explain the factors that determine the species and the number of fish to stock.
- Explain how changes in water level, water temperature, and rates of bank erosion influence the management of fish habitats.

REFERENCES

Anderson, L. G. 1977. The economic of fisheries management. The Johns Hopkins University Press, Baltimore.

Carlander, K. D. 1997. Handbook of freshwater fishery biology: Life history data on Ichthyopercid and Percid fishes of the United States and Canada. Volume Three. Iowa State University Press, Ames, Iowa.

Cole, G. A. 1994. Textbook of limnology. Waveland Press, Prospect Heights, Illinois.

Gorman, O. F., and J. R. Karr. 1978. Habitat structure and stream fish communities. Ecology 59:507–515.

Kohler, C. C., and W. A. Hubert, editors. 1993. Inland fisheries management in North America. American Fisheries Society, Bethesda, Maryland.

Nielson, L. A., and D. L. Johnson, editors. 1983. Fisheries techniques. American Fisheries Society, Bethesda, Maryland.

Wetzel, R. G. 1983. Limnology. W. B. Saunders Company, Philadelphia.

8

Outdoor Recreation Management Practices

OUTDOOR RECREATION is a voluntary, mostly *on-site* activity that people engage in for pleasure and depends on a pleasing natural setting. Viewing scenic landscapes and wildlife is also a form of outdoor recreation. The importance of wildlife to outdoor recreation in hunting for sport, viewing, and photography is paramount to many local economies. Other recreational activities include hiking, camping, sightseeing, and picnicking. Knowledge of the level of nonconsumptive use of natural resources is incomplete, although it is known that growing numbers of people are spending increasing amounts of time and money on recreational activities.

DEFINITIONS

Outdoor recreation provides relaxation and refreshment in a setting that is removed from most people's typical workplace. For the purposes of this chapter, *outdoor recreation management* provides outdoor recreational opportunities in nonurban settings. The *recreational experience* is key to how people perceive a recreational activity. The *recreational carrying capacity* is the type, amount, timing, and distribution of recreational use that is consistent with the maintenance of the ecological and social conditions specified in a management plan.

OUTDOOR RECREATIONAL ACTIVITIES

Outdoor recreation is becoming a common use of natural ecosystems in the United States (Box 8.1). If the current trend continues, future recreational values are likely to rival or eclipse those for wood and forage resource uses on many of lands.

Value of Wildlife

One of the more widely pursued outdoor recreational activities is the hunting of game animals for sport (see Chapter 6). Hunting for sport (a consumptive form of wildlife use) takes place on open public lands, outside of parks and preserves, or (with permission) on privately owned lands. Viewing and photographing wildlife (a nonconsumptive wildlife use) occur mainly in parks, preserves, and other sanctuaries where the numbers of visitors is controlled to limit disturbances of wildlife populations. These parks and preserves often become important tourist attractions for nonconsumptive outdoor recreation and, therefore, must be managed accordingly.

Box 8.1

Increasing Importance of Outdoor Recreation: One Example

Outdoor recreational use of the high-elevation, mixed conifer-ponderosa pine forests in the southwestern United States continues to increase. These diverse ecosystems provide a variety of outdoor recreational opportunities to people in the region. Most of these people live in the large metropolitan areas of Albuquerque, Phoenix, and Tucson. These city-dwellers escape the summer heat by traveling to the higher, cooler forests to participate in hunting, fishing, hiking, camping, picnicking, and sightseeing activities.

Other Outdoor Recreational Activities

People engage in a variety of other outdoor recreational activities. These activities have various impacts on the land and nat-

Table 8.1. Outdoor Recreational Activities

Activities with major impacts	Activities with minor impacts
Hunting	Hiking
Horseback riding	Camping
Off-road vehicle travel	Fishing
Second-home building	Boating
	Rock hounding
	Mountain climbing
	Relic hunting
	Bird watching
	Picnicking

Source: Holechek et al. (1998).

ural resources. Some activities are destructive to the environment and can lead to severe land and resource degradation if uncontrolled. Off-road vehicle travel is a prime example (Table 8.1). Other recreational activities have relatively minor environmental impacts.

All forms of outdoor recreation have increased in the past few decades. Much of the increase in outdoor recreational activities is attributed to people's increased affluence and a greater amount of leisure time.

RECREATIONAL EXPERIENCE

Natural resources cannot be planned nor managed to provide recreational experiences for people unless well-defined social, development, and management norms are known. *Norms* are the social standards, recreational-use preferences, and behavior patterns of people that would distinguish one group of people from another. A recreation manager needs to recognize the *recreation opportunity spectrum* to be completely responsive to society's norms. The recreation opportunity spectrum ranges from the opportunities in a pristine environment to intensively managed environments found in urban settings (Figure 8.1). The discrete segments of this spectrum constitute a particular experience. The manager must incorporate the recreational experience into the management objectives and be willing to take the necessary steps to ensure the implementation of the objectives.

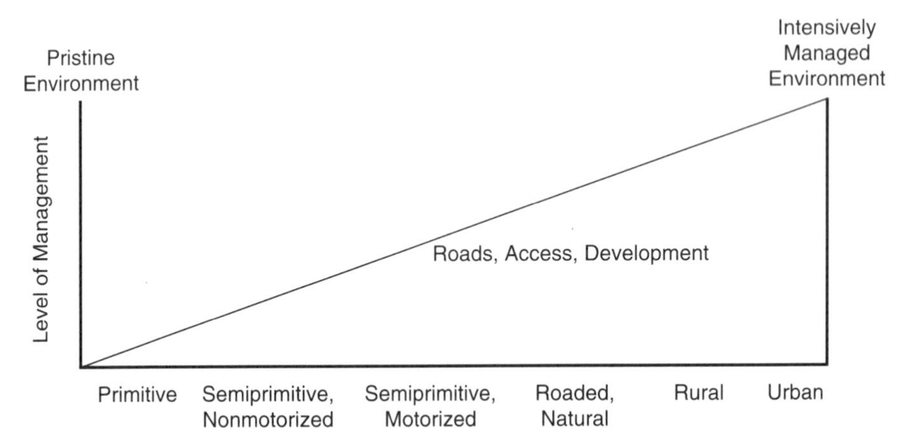

Figure 8.1. The recreation opportunities spectrum (adapted from Anonymous 1982, 1986, as presented by Becker and Jubenville 1990).

A recreational experience is often considered to be actual participation in a recreational activity. However, the actual experience is much more than this and can be separated into the following four parts:

- Anticipation—The period of waiting for a recreational activity to occur, which engages the imagination and, hopefully, develops enthusiasm. A recreational activity may never occur but still contribute to a person's happiness and satisfaction through anticipation.
- Planning—Planning for a recreational activity includes gathering equipment, obtaining supplies, packing the necessary items, and preparing other logistics. Sometimes, planning involves physical training to prepare for the recreational activity.
- Participation—A recreational activity and the events surrounding the activity extend from the time of departure to return. Participation is the heart of the recreational experience. It is the time of encounter with natural resources and activity opportunities.
- Recollection—After participation, a recreational experience is not necessarily over. Participation in the recreational activity may be relived through pictures, stories, and memories. Through time, the experience often develops additional significance and gains embellishments in the recollection phase.

RECREATIONAL CARRYING CAPACITY

The *recreational carrying capacity* concept has been useful in planning outdoor recreation management practices. Use of this concept in outdoor recreation management requires the presence of management objectives that measure the recreational setting in terms of:

- The *physical-biological environment,* for example, the level of vegetative disturbance (if this is relevant to the recreational setting) or water quality characteristics.
- The *social environment,* that is, the types of activities allowed or levels of interaction among visitors necessary for the recreational experience to be satisfactorily attained and maintained.

From a manager's perspective, the application of the recreational carrying capacity concept is a resource allocation process, where people achieve the specified recreational opportunity through the allocation of:

- Land, water, and other natural resources for recreational purposes.
- Physical developments, services, natural resource management programs, regulations, and other fiscal resources for recreation purposes.

OUTDOOR RECREATION MANAGEMENT PRACTICES

Outdoor recreation management practices are separated into three categories: site management, visitor management, and service management. While there is overlap among these management categories, their separation is useful for planning purposes. Managers of outdoor recreation should be aware that these categories always involve relationships between people and natural resources.

Site Management Practices

Outdoor recreational sites can be classified into sites where the impacts of recreation occur at points of concentrated use (campgrounds and picnicking areas or boat launching ramps); sites with linear areas of impact (trails or roads); and sites with areas of dispersed impact (off-road vehicle areas or trailless hiking areas). Regardless of the nature of the recreational site, the most effective way to avoid unacceptable impacts involves careful site selection, facility design, and a maintenance plan to minimize impacts. Avoiding impacts does not necessarily involve reducing site use.

Site Selection

A key element of outdoor recreation management is site selection. Among the factors considered when selecting a site for a recreational use are the soil and vegetation, durability of the site to withstand the recreation use, pollution potential, and potential for adverse impacts on wildlife. Other factors include landform, aspect, elevation, and proximity to water. Many impact problems are attributed to locating recreational use areas on sites that are too moist or subject to severe erosion.

When off-road vehicle areas are located on steep terrain, unacceptable impacts are inevitable. Alpine tundra, moist meadows, sensitive bogs, riparian communities, and critical wildlife habitats should also be avoided for this recreational activity. Areas that are designated for off-road vehicle use should be located where off-site impacts (such as downstream sedimentation) are minimized.

Trails. Providing comfortable pathways for walking encourages visitors to remain on trails. Dense vegetation, obstacles, such as logs and rocks, and purposely routing hikers through rough terrain all serve to confine the activities of hikers. Heavily used trails should be hardened when compatible with other management objectives. Wet sites should be drained or bridged with stepping-stones or rock-treadways. Trails should have switchbacks or rock steps when traversing steep slopes. Streams should not be diverted nor should streamflow be interrupted when trails are designed. Culverts used for crossing a stream must be large enough to accommodate the highest flood-stage streamflow. Stream crossings without bridges should be approached at a downstream angle to minimize trail erosion.

Camping and Picnicking Areas. Camping and picnicking areas should be concentrated on durable sites and features that are likely to promote use should be provided, such as tables and benches, fire rings, and tent pads. Use on these sites needs to be channeled along well-defined routes and concentrated in activity centers. The distance between sites and service facilities should be minimized and trails should link the facilities. Activity centers and pathways should be hardened where compatible with other management objectives.

Picnic tables need to be fastened to the ground or to be so heavy that they cannot be moved about the site. Fires should be confined to one part of the site so that the impacts from campfires do not spread. Either firewood should be provided or visitors should be directed to firewood collection areas. Natural barriers (vegetation, rocks, or earth) or constructed barriers are often used to confine and direct vehicular and pedestrian traffic. Some sites designed to accommodate large parties without regulations on party size should be available. The general facility design should exclude the possibility of overuse by limiting parking lot size or in the case of backcountry recreation, reducing accessibility.

Visitor Management Practices

Indirect and direct techniques are employed in visitor management. *Indirect techniques* attempt to modify the factors that influence visitor decisions and behavior without (in most instances) visitors perceiving the control of their activities. *Direct techniques* modify visitor decisions and behavior with visitors usually being aware that their activities are controlled. Indirect techniques should be used before direct techniques whenever possible.

Indirect Techniques

Because they permit visitors the freedom to choose, indirect techniques are considered to be more consistent with the free and pleasurable characteristics of a recreational experience than are direct techniques. Some indirect techniques improve visitors' experiences by providing useful information to help them locate what they are seeking. Indirect techniques are usually less costly than direct regulation and accomplish the same goal, but with less hassle. Also, there are relatively few problems with rule violations with indirect techniques. Recreation managers should not be placed in a confrontational role; rather, managers should be help-

Table 8.2. Indirect Management Techniques

Methods	Techniques
Physical alterations	Improve, maintain, or neglect access roads
	Close selected roads
	Build new roads
	Modify parking facilities
	Make trails more or less difficult
	Build trails or leave areas trailless
	Modify access to water (near boat ramps, near trails, or none)
	Improve fish or wildlife populations, or not
	Create water bodies
	Open views
Information distribution	Provide more or fewer directional signs
	Use messages (signs, publications, etc.) to advertise areas and opportunities
	Inform visitors of use patterns (time of use or maximum or minimum use) particularly for underused areas
	Educate users in desired behavior
Eligibility requirements	Charge constant fees
	Charge fees that vary among areas and seasons

ful facilitators. Examples of the indirect management techniques are in Table 8.2.

Direct Techniques

Indirect techniques alone are not always effective to achieve a desired recreational experience and may need to be supplemented by direct techniques. Sometimes to assure safety, certain regulations are required. However, regulations can also present problems if they are not effectively communicated to visitors. Visitors should be aware of regulations, know clearly what is required of them, and have some understanding of the reasons for the creation of the regulations. Enforcement of regulations needs to create a situation that is fair for those who comply because compliance often requires some inconvenience. Some examples of direct management techniques are included in Table 8.3.

Service Management Practices

Service management practices provide the necessary facilities, services, and access to visitors. Service management is crucial

Table 8.3. Direct Management Techniques

Methods	Techniques
Enforcement of regulations	Increase patrolling
	Impose fines
	Hire watchmen
	Use surveillance televisions or cameras
Zoning	Prohibit some types of use in certain zones (for example, no motorboats in some waters or prohibit some types of use at certain times)
	Close some campsites (for example, near lakes)
Regulate or ration use	Limit the number of persons admitted
	Limit the group size
	Limit the length of stay
Restrict activities	Prohibit campfires
	Prohibit off-road travel or trails
	Prohibit camping
	Prohibit dogs or require leashes
	Prohibit littering
	Prohibit grazing by visitors' horses, etc.

when visitors must travel a long way from home to enjoy a recreational setting. Service management begins with recreation-area planning and involves concession, maintenance, and hazard management.

Recreation-Area Planning

Recreation-area planning is an allocation process in which natural (land, water, and air) and fiscal resources (money for facilities, services, and access) are allocated in a manner that provides optimal recreational opportunities to visitors. A result of this planning effort is the recreation-area plan, which is the vehicle used to accomplish the task of providing outdoor recreation as a land-use activity.

The facilities, services, and access provided to the visitors are human-made attributes. Visitors respond by selecting a recreation site. The recreation-area planning process should identify the need for concessions, maintenance, and hazard reduction.

Concession Management Practices

Concession management includes planning, developing, and supervising concession services to enhance the quality of the recreational experience. Recreation managers should initially determine what services are needed and where they should be located. Common concessions are food services, transportation systems, gift shops, and overnight accommodations (other than on-site camping facilities). The specific location of concessions must be identified in the recreation-area plan. If a recreational area is publicly owned, it must be determined whether the concessions will be owned and operated by the government, by people in the private sector, or a combination of the two. Concession leasing arrangements or special-use permits, which specify the relationship between a concessionaire and a government agency, are often necessary.

Short- and long-term environmental impacts of concessions must be assessed. Mitigating measures should be developed to alleviate environmental problems or relocate concessions to reduce detrimental environmental impacts. Supervisory responsibilities should be known for efficient concession operations and to ensure that the concessions furnish the quality services necessary to enhance the visitors' experience.

Maintenance Management

Maintenance management refers to the care required to ensure that facilities, services, and access on a recreational site furnish a continuing, high-quality, and safe visitor experience. Maintenance management includes the efforts necessary to maintain trails and camping and picnicking areas at established standards.

Trails. Trail maintenance entails the removal of obstacles (trees, rocks, and slides) and the repair of water bars and other drainage control devices. Maintenance usually begins before the start of the main-use season. Trails not meeting the established standards are hardened, or when necessary additional drainage control devices are installed. Trails that need to be relocated because of overuse should not be moved unless visitors can be kept from using the trail and a more durable route is available.

Camping and Picnicking Areas. Maintenance of camping and picnicking areas requires continuous monitoring of the areas to ensure established standards are met. Areas can remain open for use during maintenance or can be periodically closed. However, periodic closure of camping and picnicking areas can be self-defeating when site use is heavy. The recovery periods are usually long when compared to the time it takes for unwanted impacts to occur. Camping and picnicking activities should be relocated to more durable sites or visitor use greatly reduced when the sites cannot be maintained without periodic closure.

Maintenance management also entails the maintenance of outhouses, solid-waste disposal facilities, bridges, visitor centers, and clean water supplies. All maintenance management must be conducted at a standard that meets visitors' expectations, agency policies, and safety requirements.

Hazard Management

Hazard management is aimed at reducing the probability of injury, loss of life, or loss of property from natural and human-made hazards. Natural hazards include overhanging cliffs, landslides or falling rocks, lightning, flooding, and potentially threatening wildlife, such as bears, moose, or rabid coyotes. Poorly maintained or abandoned buildings, vehicle bottlenecks and blind turns on roadways, and altered landscapes, such as wells and mine shafts, are examples of human-made hazards. It must always be remembered that visitors are still in charge of their own behavior and actions, but they can minimize the risks to themselves when they comply with hazard management programs.

Outdoor recreation managers plan and offer hazard management programs to warn visitors of potential dangers. Techniques commonly used in these programs include site planning, information and education, and establishing regulations.

VISUAL RESOURCES MANAGEMENT

Planning and management for the use of natural resources is becoming increasingly complex. The most vexing problem concerns how best to integrate traditional economic values (forage, timber, and water) with less tangible values (outdoor recreation and viewing scenic landscapes). While some people argue that these intangibles can never be properly quantified, others believe that there are opportunities for incorporating them into the planning and management processes.

Visual resources management is the management of the "seen aspects" of landscapes and the activities that occur on these landscapes. Visual resources maintain or upgrade people's psychological welfare and their recreational experience. Providing for the viewing of a scenic landscape is included in visual resources management. Visual resource management involves:

- Inventorying and analyzing the landscape with one of the many systems designed for this purpose, for example, either the USDA Forest Service or the Bureau of Land Management systems[1], or one of the other approaches for estimating the value of scenic beauty (Box 8.2).
- Setting visual quality objectives, which may include preservation, retention, partial retention, or modification of the landscape.
- Incorporating all of the above and other resource information into a land-use planning process.
- Post-evaluating the plan and the management accomplishments concerning the visual resource.

Box 8.2

The Scenic Beauty Estimation Method (Daniel and Boster 1976, Brown and Daniel 1984)

The scenic beauty estimation (SBE) method is based on psychological and psychophysical measurement theories (Thurstone 1959), categorical scaling models (Torgerson 1958, Nunnally 1978), and principles from signal detection theory (Green and Swetts 1966). In applying the SEB method, landscape scenes are represented by a systematic photographic survey, that is, a number of randomly located, randomly oriented color slides. These photos are presented to observers who independently rate each scene on a ten-point scale, with ten being the highest value. The ratings are transformed following Thurstone's procedures and guidelines from signal detection theory, and an interval

[1]National Forest Management. The visual management system. USDA Agriculture Handbook 462, Washington, D.C.

Visual Resource Management. USDI Bureau of Land Management Manual 8400, Washington, D.C.

scale index of perceived scenic beauty, the scenic beauty estimator, is obtained. The ratings are adjusted to decrease the effect of extreme scores (very high or very low values) by using a scaling procedure that presents SEB values as an unbiased measure of differences in perceived scenic beauty.

Visual resources should not be managed separately from other resources in the decision-making process. However, natural ecosystems offer a multitude of opportunities for leisure activities, all of which are enhanced by the scenic beauty of the landscape. This requires that appropriate visual resources management be an active component of the multiple-use, ecosystem-based management of natural resources.

SUMMARY

Outdoor recreational activities are becoming a major component of natural resources management. People are demanding more opportunities for hiking, camping, picnicking, hunting, fishing, and sightseeing. The challenge confronting recreation managers is carefully integrating site, visitor, and service management practices to ensure that people's desired recreational experiences can be met within the limitations of the physical, biological, and social environments.

After reading this chapter, you should appreciate and understand:

- What is meant by the recreational experience, including the meanings of anticipation, planning, participation, and recollection.
- The role of management practices to modify the factors that indirectly influence visitor decisions and behavior, and the management practices that directly modify visitor decisions themselves.
- The role of site selection, facility design, and maintenance in the management of sites where outdoor recreation takes place.
- The importance of site, visitor, and service management practices to the recreational experience.

- The necessity of providing for the viewing of scenic landscapes through visual resources management.

REFERENCES

Anonymous. 1982. R.O.S. users guide. USDA Forest Service, Washington, D.C.

Anonymous. 1986. R.O.S. book. USDA Forest Service, Washington, D.C.

Becker, R. H., and A. Jubenville. 1990. Forest recreation management. In: Young, R. A., and R. L. Giese, editors. Introduction to forest science. John Wiley & Sons, Inc., New York, pp. 404–426.

Brown, T. C., and T. C. Daniel. 1984. Modeling forest scenic beauty: Concepts and application to ponderosa pine. USDA Forest Service, Research Paper RM-256.

Daniel, T. C., and R. S. Boster. 1976. Measuring landscape esthetics: The scenic beauty estimation method. USDA Forest Service, Research Paper RM-167.

Green D. M., and J. A. Swetts. 1966. Signal detection theory and psychophysics. John Wiley & Sons, Inc., New York.

Hammitt, W. E., and D. N. Cole. 1998. Wildland recreation: Ecology and management. John Wiley & Sons, Inc., New York.

Holechek, J. L., R. D. Pieper, and C. H. Herbel. 1998. Range management: Principles and practices. Prentice Hall, Upper Saddle River, New Jersey.

Jubenville, A., and B. W. Twight. 1993. Outdoor recreation management: Theory and application. Venture Publishing, State College, Pennsylvania.

Malgill, A. W. 1992. Managed and natural landscapes: What do people like? USDA Forest Service, Research Paper PSW-RP-213.

Nunnally, J. C. 1978. Psychometric theory. McGraw-Hill Book Company, New York.

Thurstone, L. L. 1959. The measurement of values. University of Chicago Press, Chicago.

Torgerson, W. S. 1958. Theory and methods of scaling. John Wiley & Sons, Inc., New York.

Wagner, J. A. 1964. The carrying capacity of wild lands for recreation. Forest Science Monograph 7, Society of American Foresters, Washington, D.C.

Zinser, C. 1995. Outdoor recreation: United States national parks, forests and public lands. John Wiley & Sons, Inc., New York.

9

Wilderness Management Practices

THE SPECTRUM of outdoor recreational opportunities is represented by a continuum that begins with opportunities on totally undeveloped lands and pristine areas to the heavily developed urban landscapes discussed in the preceding chapter (Figure 8.1). The focus of this chapter is placed on the primitive end of that spectrum.

DEFINITIONS

A *primitive area* is land that has been set aside for the preservation of natural conditions and little or no alterations or development is permitted, with the exception of measures needed for fire protection. With the exception of some legal differences, a *wilderness area* is similarly defined. The general term *wilderness* will be used for the discussion of the management practices included in this chapter.

Wilderness has many definitions. However, Aldo Leopold, a noted environmentalist, wildlife manager, and the author of *A Sand County Almanac,* proposed a definition that serves the purpose of this book. While employed by the USDA Forest Service in the Gila River Basin of New Mexico, Leopold defined *wilderness* as "a continuous stretch of country preserved in its natural state, open to hunting and fishing, big enough to absorb a two-week pack trip, and kept devoid of roads, trails, homes, and other works of people." To date, this definition of wilderness has remained largely unchanged.

Wilderness management provides outdoor recreational opportunities in a wilderness setting. *Wilderness management practices* are carried out to furnish people with opportunities to enjoy a wilderness experience, while attempting to maintain the pristine character of the wilderness environment.

WILDERNESS MANAGEMENT PRINCIPLES

The following principles offer the wilderness manager a perspective on the unique nature of the wilderness resource, its appropriate use, and its place in the spectrum of land-use practices. These principles also provide a framework for formulating wilderness management practices. The principles are:

- Wilderness represents one extreme on the recreational opportunity spectrum.
- Wilderness is a distinct natural resource with inseparable parts.
- Wilderness preservation requires the management of people's use of the wilderness and their impacts on an area.
- Wilderness preservation also requires recognition and consideration of the recreational carrying capacity concept (see Chapter 8).
- Wilderness management practices must be guided by explicit objectives set forth in a comprehensive plan of action.
- Wilderness management should focus on producing wilderness-dependent values and benefits to people.
- Wilderness management must be viewed holistically in relation to the management of adjacent landscapes.
- Wilderness-dependent activities should be favored in managing wilderness use.
- Wilderness management should strive to selectively reduce the physical and social-psychological impacts of use.
- Only the minimum regulations necessary to achieve the stated objectives of wilderness management should be applied.
- Management of individual wilderness areas should be largely governed by a concept of nondegradation.

WILDERNESS EXPERIENCE

Wilderness management practices should provide a set of human experiences that are linked to the naturalness of the wilderness setting (Box 9.1) and, therefore, to a very low level of development and contact with other people. These landscape features are the "best attributes" of the *wilderness experience* for many people. The qualities of the naturalness and solitude of the wilderness setting are subject to change, however. Opportunities for solitude decline and the probability for substantial modifications of a wilderness ecosystem increases as outdoor recreational use increases. Excessive increases of recreational use are reflected by increases in social costs, environmental costs, or a combination of the two. A loss in the carrying capacity of a wilderness can follow, primarily because increases in outdoor recreational use can adversely affect the long-term capability of the resource to produce the wilderness-dependent experiences that are desired.

Box 9.1

Activities Contributing to Wilderness Experiences

Wilderness experiences are comprised of a variety of outdoor recreational activities. Land-based activities in wilderness areas include the viewing of scenery, hiking and horseback riding (but not on developed trails), camping (but not on developed sites), hunting, nature study, and mountain climbing. Water-based activities are canoeing and the use of other nonmotorized watercraft, swimming, and fishing. Management of wilderness areas furnishes opportunities for these and other outdoor recreational activities that satisfy people's expectations.

WILDERNESS MANAGEMENT PRACTICES

The categories of management practices outlined for outdoor recreation in Chapter 8 are similar to those used for wilderness management practices, which include site management, visitor management, and service management. Wilderness managers must realize that the relationship between people and the unique natural resources of wilderness areas must guide these management practices.

Site Management Practices

Trails and campsites are acceptable in wilderness areas and are compatible with the wilderness concept. Trails and campsites are indispensable to the use of many wilderness areas. However, management objectives for wilderness areas must reflect a proper balance between the need for trails and campsites and the natural environments that people desire. Trails and campsites should be managed to minimize the impacts of use while still providing the opportunity for high-quality wilderness experiences.

Trails

Trails are the main travel arteries in wilderness areas. Unfortunately, few wilderness trails were originally built with either recreation or wilderness protection in mind. They were mainly built to provide access for fire control. Improvements of many trails could help to reduce resource impacts. Design standards for trail improvement should specify the following: pathway width, grade (slope), alignment, surface materials, drainage, amount of trailside vegetation to clear, extent to which down trees or other obstacles should be cleared, signs, and types and locations of bridges. Visitor surveys often indicate little support for high-standard trails and a high level of support for low-standard trails. This is a positive trend because in the past wilderness trails have often been built to excessively high standards.

Campsites

Campsites can be an integral component of wilderness management. The number of campsites that offer seclusion (that is, out of sight and sound of other campsites) is often a major determinant of wilderness carrying capacity. However, visitors tend to congregate at campsites, sometimes in large numbers for long periods of time, which results in drastic alteration of extensive areas. Therefore, management decisions rely on the availability of established and potential campsites and campsite location, attraction, and condition.

Opinions differ widely on where campsites should be located relative to the location of trails, lakes, and streams. Some wilderness managers prohibit camping within a specified distance of trails. Other managers prohibit camping at sites close to lakes or streams. Reasons for restricting streamside or lakeside camping include preventing damage of fragile vegetation, decreasing the possibility of polluting the water, and making environmental damage less noticeable. Nevertheless, there is overwhelming evidence that the most popular campsites are close to water. No-camping zones are sometimes established to eliminate problems. However, care must be exercised that no-camping zones do not eliminate all the campsites located close to the water.

The type of camping facilities that should be provided to people in a wilderness area is another controversial topic. There are varying viewpoints on the number of different facilities that is justified at a campsite. Facilities that are often found at wilderness campsites include fireplaces, tables, tent pads, shelters, horse facilities, toilets, garbage pits, and water pumps. The camping facilities that visitors expect to find at a site must be determined and then provided.

Visitor Management Practices

Wilderness management primarily centers on the management of visitors. A framework for visitor management in wilderness areas of the United States is derived from the wording in the Wilderness Act—wilderness is "underdeveloped . . . land retaining its primeval character and influence, without permanent improvements or human habitation . . . ." When considering those aspects of visitor actions in a wilderness that require a management response, it is helpful to think of the responses listed in Table 9.1:

Table 9.1. Wilderness Visitor Actions and Management Responses

Type of Visitor Action	Response
Illegal	Law enforcement
Careless	Persuasion, education, rule enforcement
Unskilled	Education, rule enforcement
Uninformed	Education, information
Unavoidable	Reduction in use, relocation of use to other sites

Wilderness managers generally mitigate the impacts of visitor actions or control usage levels through application of one or more of the following management tools:

- Amount of use.
- Timing of use.
- Party size.
- Length of stay.
- Visitor behavior in relation to:
  - Resource impacts.
  - Impacts on other visitors.
  - Cooperation with management.

Management of wilderness area visitors has often been based on the manager's personal experiences, feelings, and intuition rather than on the systematic and objective collection of information. However, the professional judgment of a wilderness manager should always be centered on reliable information about use, existing resource condition, likely user responses to management actions, and probable responses of the wilderness ecosystem to management decisions.

Service Management Practices

Service management in wilderness areas is mainly concerned with trail maintenance and the restoration of campsites. Trail maintenance can be a major task of the wilderness manager, with the biggest job being removal of trees and shrubs that fall across the trails. Trail-maintenance crews also prune encroaching branches and clear large rocks from the trail; repair erosion damage; and install or maintain bridges, ditches, drains, culverts, and water bars.

Restoration of campsites to their natural condition is a frequent task confronting wilderness managers. Almost all actions taken to restore damaged campsites involve a temporary closure of the campsite to provide time for rest and recovery. However, there is a range of restoration actions that can accompany closures, including:

- Rest—No other action.
- Rest—Seedbed preparation, natural regeneration.
- Rest—Seeding and planting.

- Irrigation.
- Fertilizing.
- Eradication of nonnative plants.
- A combination of these restoration actions.

The challenge for managers is selecting the best action for a particular wilderness situation. The effectiveness and impacts of each of the possible restoration actions must be evaluated in relation to site-specific conditions.

FIRE ISSUES

An important issue for a wilderness manager to address is whether to exclude fire from the ecosystem. Much of the wilderness landscape in the United States has been shaped by repeated occurrences of wildfire. However, in many natural ecosystems fire has been excluded or controlled since advent of the *Smokey the Bear* campaign in the 1940s, which pointed out the damaging aspects of fire. The dilemma is choosing the best policy to follow with respect to the likeliness for fires to occur, both lightning- and human-caused, and current wilderness management practices. In general, there are five options:

- Total fire exclusion.
- No fire control program.
- Management of lightning-caused fires that burn within the limits of a fire prescription, that is, *prescribed natural fires* (see Chapter 10).
- Restoration of a natural fire regime through the substitution of prescribed fire for lightning-caused fire.
- Exclusion of fire and the mechanical manipulation of vegetation and fuels to achieve desired conditions.

Which option is accepted and implemented is normally resolved on a case-by-case basis. If the goal of wilderness management is maintaining an environment where natural fires are allowed to burn within prescribed conditions, a policy of fire suppression could be modified to allow for an active program of prescribed burning. However, for some people this policy violates the spirit and intent of wilderness preservation (Box 9.2). Regardless of the action taken, excessive damage by fire must not be allowed to impact the natural function of the ecosystem, cultural resources, or personal property on adjacent lands.

Box 9.2

Fire Polices and Programs: A U.S. National Park Service Perspective (Kilgore and Nichols 1995)

In their efforts to balance caution with reason in the aftermath of the Yellowstone fires of 1988, personnel of the U.S. National Park Service concluded that they did not need to suppress all "ecological significant fires in park and wilderness ecosystems." Fire has always been a component of most park and wilderness ecosystems. Park service personnel recommended polices that allowed managers to restore fire to an ecosystem when the land management objective is perpetuating natural processes and refining, strengthening, and reaffirming values. Furthermore, overreaction to the events of 1988 should not be used to justify curtailing prescribed natural fire.

SUMMARY

Wilderness management practices are undertaken to ensure sustainable "primitive" lands and to provide human experiences that are linked to these wilderness settings. Similar to the goals of an outdoor recreation manager, a wilderness manager must keep in mind that the relationship between people and natural resources is a key to wilderness management.

After reading this chapter, you should be able to:

- Explain what is meant by the term *wilderness,* and what are the features of a wilderness.
- Discuss the management principles that place the wilderness resource in its proper place on the spectrum of land-use practices.
- Describe the management tools available to mitigate the impacts of visitor actions on wilderness areas.
- Understand how trails and campsites can be managed to ensure

the wilderness experience. Explain how these sites can be restored.

- Discuss the options available to a manager regarding the occurrence of natural fires in wilderness areas.

REFERENCES

Cole, D. N. 1994. The wilderness threats matrix: A framework for assessing impacts. USDA Forest Service INT-475.

Fischer, W. C. 1984. Wilderness fire management planning guide. USDA Forest Service, General Technical Report INT-171.

Hendee, J. C., G. H. Stankey, and R. C. Lucas. 1978. Wilderness management. USDA Forest Service, Miscellaneous Publication 1365, Washington, D.C.

Hammitt, W. E., and D. N. Cole. 1987. Wildland recreation: Ecology and management. John Wiley & Sons, Inc., New York.

Kilgore, B. 1985. What is "natural" in wilderness fire management? USDA Forest Service, General Technical Report INT-182.

Kilgore, B. M., and T. Nichols. 1995. National Park Service fire policies and programs. In: Brown, J. K., R. W. Mutch, C. W. Spoon, and R. H. Wakimoto, technical coordinators. Proceedings: Symposium on fire in wilderness and park management. USDA Forest Service, General Technical Report INT-GTR-320, pp. 24–29.

Landres, P., and S. Meyer. 1998. National wilderness database: Key attributes and trends, 1996 through 1998. USDA Forest Service, General Technical Report RMRS-GTR-18.

Leopold, A. 1949. A Sand County almanac and sketches here and there. Oxford University Press, New York.

Stankey, G. H., D. N. Cole, R. C. Lucas, M. E. Petersen, and S. S. Frissell. 1985. The limits of acceptable change (LAC) system for wilderness planning. USDA Forest Service, General Technical Report INT-176.

10

Fire and Pest Management Practices

FIRE, INSECT pests, and diseases are part of natural environments. Little can be done to totally exclude their occurrences, even if such exclusion was desirable. However, natural ecosystems must be protected from the detrimental effects of large-scale wildfire and insect and disease epidemics to maintain a healthy, productive status. Strategies and management practices for protecting natural ecosystems from fire and pests are presented in this chapter.

DEFINITIONS

Fire is the rapid release of the heat energy stored in plants by photosynthesis. The mention of fire often evokes images of flames. However, there is more to fire than flames. Fire is the manifestation of a series of chemical reactions, while flames are a gas-phase phenomenon of fire; therefore, they are only part of the process.

A *natural fire* is a fire of natural origin, often the result of lightning. A *human-induced fire* is a fire caused directly or indirectly by a person or people. A *wildfire* is a fire that is not meeting management objectives and requires suppression. A *prescribed fire* is a fire resulting from a planned human-ignition within a defined prescription. The prescription includes a statement specifying the management objective for setting the fire and the allowable weather conditions, fuel moisture condition, and soil moisture condition under which the fire will burn. A *prescribed natural fire* is a fire of natural origin that is allowed to burn as long as it is accomplishing stated management objectives.

Pests are insects, diseases, and other damaging agents that are detrimental to ecosystem health. *Integrated pest management* combines chemical, biological, cultural, and mechanical practices to maintain pest levels below a preestablished threshold. Integrated pest management often is a key to pest control when other control methods fail. *Ecosystem health* is a condition where an ecosystem is capable of recovering from a wide range of disturbances and retaining its ecological resiliency, while meeting people's current and future needs for desired values, uses, products, and services.

FIRE

Fire is a threat to all natural ecosystems. Even in relatively high rainfall areas, there can be hot, dry periods when the risk of wildfire is high. Initially a wildfire generally burns at the ground level. Once the fire rapidly spreads on the ground and increases in intensity, the wildfire can move into the crowns of trees and shrubs, causing greater damage. Therefore, the risk of wildfire is a major concern in planning for the sustainable management of natural resources.

Wildfire originates from natural causes (lightning strikes or spontaneous combustion) or the actions of people. The latter often results from fires purposely started that unintentionally spread beyond a defined perimeter. These types of fires include fires set on fallow agricultural croplands; fires set on rangelands for the improvement of livestock grazing conditions; and the fires associated with the activities of hunters and recreationalists. There are also instances of deliberate burning, which can be motivated by the creation of employment opportunities from fire suppression activities and the subsequent replanting of fire damaged areas; or to show disapproval of natural resources management policies.

Prevention

It is generally not possible to prevent natural buildup of combustible materials (fuel), which create conditions favorable for fire occurrence. However, the use of fire prevention campaigns can minimize the risk of fire caused by people. The most visible prevention campaigns are educational programs that use radio messages, signs and message boards, news releases, and magazine articles. In the United States, programs such as *Smoky the Bear* and *Keep America Green* have relied heavily upon public education.

Knowing the cause of fires helps direct the thrust of prevention campaigns. For example, the high frequency of fires along roadways where most cigarette-caused fires occur has led to regular appeals to crush cigarettes and to use car ashtrays. A high incidence of fires attributable to escaped campfires can indicate a need to install outdoor fireplaces. Arson is a more difficult problem to address, although it is likely that education has some effect.

Fire prevention can also be accomplished by *hazard reduction activities*. There is little risk of fire where there is insufficient buildup of combustible materials, which allow ground fires to develop. Therefore, one approach to protecting natural ecosystems against fire is removal of excessive buildup of fuels that accumulate on the ground. Prescribed burning (see below) is sometimes used to reduce accumulations of hazardous fuels, for example, in areas with heavy logging residues, along railroad tracks, and in forests where an excess level of organic litter has accumulated. Other approaches to reducing the amounts of available fuel include salvage harvesting of standing dead trees; reducing the amount of logging residues left on-site by increasing timber utilization; and mechanical manipulation of fuels by crushing or rearranging.

Constructing barriers to prevent the spread of a fire by clearing potentially combustible fuels is another form of prevention. *Firebreaks* established at specified intervals represent an example of such barriers. However, constructed barriers must be continuously maintained to ensure their effectiveness.

Fire-Danger Rating

The USDA Forest Service's National Fire-Danger Rating System is helpful in identifying the days when the danger of fire is greatest. This fire-danger rating system integrates the principal factors influencing fire potential and behavior, fuels, weather, and topography, into a burning index that represents the degree of fire danger (Figure 10.1). Burning indices are derived from three fire behavior components: the rate of spread, energy-release, and igni-

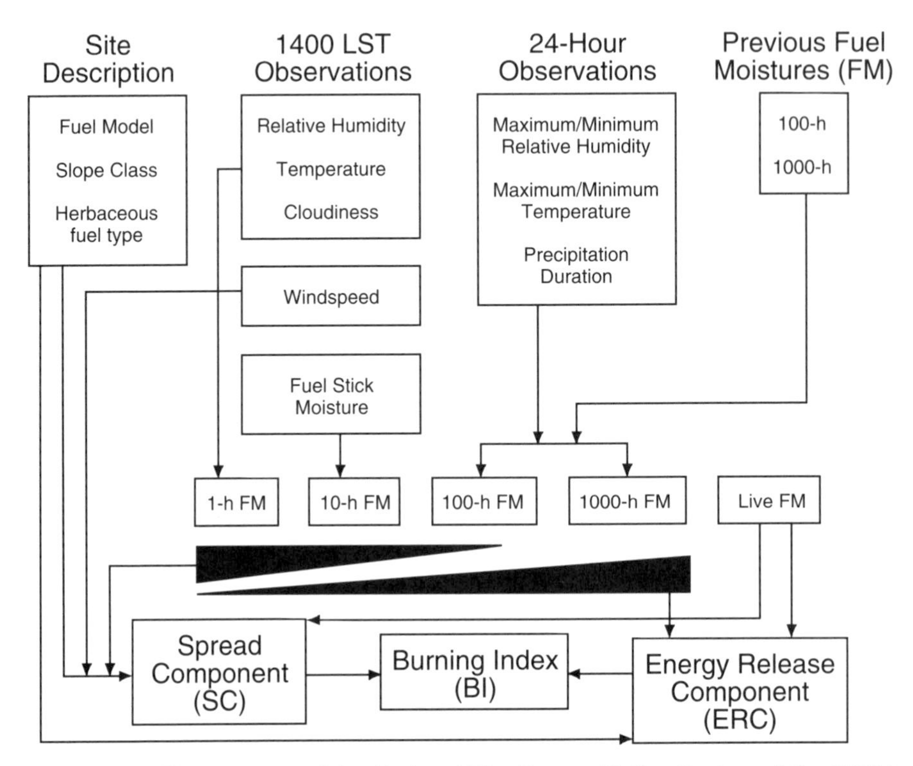

Figure 10.1. The structure of the National Fire-Danger Rating System of the USDA Forest Service showing the relationships among the site, weather observations, fuel moisture, and calculated index (from Andrews and Bradshaw (1992), as presented by Pyne et al. 1996).

[A description of how to apply the National Fire-Danger Rating System is presented in Pyne et al. 1996].

tion. Each of the indices is evaluated on a scale of zero to 100. The indices are:

- Occurrence Index (OI)—OI values indicate the potential for fires to occur within the rating area. A high OI indicates full readiness for the detection system.
- Burning Index (BI)—BI values indicate the potential amount of effort that is needed to contain a particular type of fire within a rating area. A high BI requires a large amount of labor and equipment to combat a single fire.
- Fire Load Index (FLI)—FLI values indicate the total amount of effort required to contain all probable fires occurring within a rat-

ing area for a specified period of time. Readiness plans are keyed into the FLI.

Application of the National Fire-Danger Rating System is relatively complex. Recording and observing weather, fuel accumulations, and topography using specified instruments and outlined procedures are required. These recordings will serve as a basis for the series of calculations used for obtaining the index values that comprise the danger rating system. Details for applying the system are found in Demming et al. (1977), Martin et al. (1979), Bradshaw et al. (1983), Burgan (1988), Pyne (1996), and DeBano et al. (1998).

Control Practices

Even the best fire prevention measures cannot totally eliminate the occurrence of fires. Therefore, fire control practices, including fire detection and suppression activities, become necessary.

Fire Detection

Lookout towers were the basis for fire detection in the early years of fire control in the United States. Lookout towers were placed on high land points to provide visual coverage of the landscape. The location of a fire was estimated by intersection or triangulation of sightings made from two or more lookouts. Aerial detection of fires has gradually replaced lookouts in more recent years because it is cheaper than maintaining a system of lookout towers and allows for more complete and detailed coverage of fire-prone areas. Nevertheless, many fire control agencies continue to use a skeletal system of lookouts to supplement aerial reconnaissance.

More recent developments in fire detection include the aerial use of infrared sensing equipment, which photographically records the location of heat sources. Although there are some problems with accurately identifying heat sources, the use of infrared sensing equipment in aerial reconnaissance has shown that it is possible to detect the presence of small fires, such as glowing buckets of charcoal beneath a forest canopy. This technique has the potential for detecting small and smoldering fires that do not produce enough smoke to be detected by other methods.

Suppression Activities

The three components of combustion—heat, oxygen, and fuel—are often pictured as a triangle. These three components are used for planning fire suppression activities (Figure 10.2). All three components of the triangle must be present for ignition and subsequent burning to occur. A fire fighter's task is breaking up this combination by one or more of the following actions:

- Removing the fuel.
- Reducing or removing the supply of oxygen.
- Reducing the temperature below the kindling point of the fuel.

Fuels are removed from a burning site to halt the progress of an advancing fire by digging, scraping, plowing, or otherwise clearing a strip of earth known as a *fire-line.* Applications of soil, water, or fire-retardant chemicals to the fire serve to restrict the supply of oxygen and lower fuel temperatures.

The preferred method of fire suppression is often a *direct attack.* In a direct attack, a fire-line is constructed near the fire edge and the flames are knocked-down with the use of water, soil, or other means. The fire-line can be situated at a distance if the fire cannot be approached at close range. Alternatively, a *backfire* can be start-

Figure 10.2. The fire triangle of heat, oxygen, and fuel. These three components must be present for a fire to ignite.

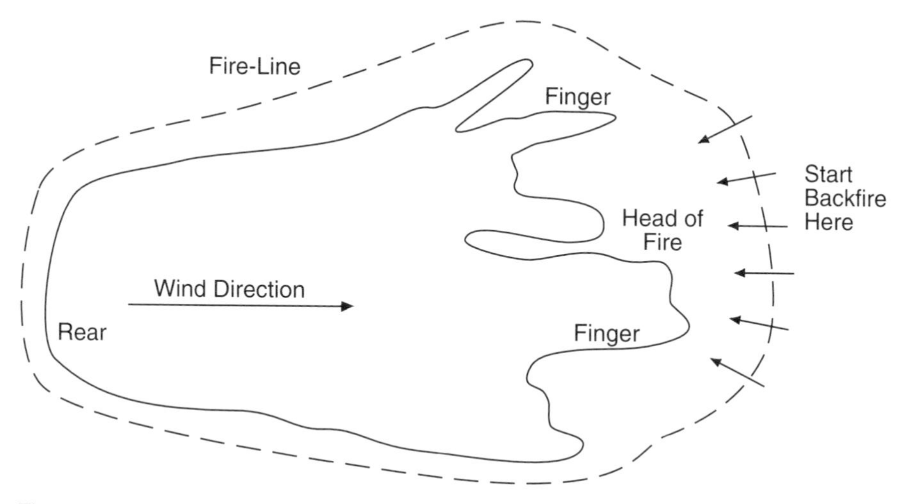

Figure 10.3. The parts of a fire with an illustration of the placement of a fire-line and backfire for suppressing a fire (from Stoddard and Stoddard 1987).

ed in front of the advancing fire and allowed to spread toward it, thus eliminating the fuel available to the advancing fire (Figure 10.3). This method of fire suppression is known as an *indirect attack.*

Suppression of large wildfires can require the use of hand tools, bulldozers, and water pumps for hours or days. Another fire suppression activity is the use of small airplanes that apply water or fire-retardant chemicals to active fire sites. Adding flame-inhibiting chemicals or other additives augments the effects of water by increasing its ability to smother burning fuels. Fine clays are often mixed with water to increase the cohesiveness of the mixture and prevent excessive scattering and evaporation during descent. These mixtures of clay and liquids are called *slurries.*

Application of slurries from the air is a helpful tactic to delay the spread of a fire, providing the time needed for people and equipment to arrive on the scene. This is important when the fire occurs in a remote location. However, retardants and slurries applied from the air are not usually sufficient to totally extinguish a fire. Follow-up work by ground crews is almost always necessary.

A fire is considered *contained* when the fire-line around the fire has been completed and the fire is no longer spreading. However, there remains the often long and tedious process of *mopping-up* after the fire. Mopping-up is necessary to ensure that the fire will not flare up. All smoldering fires that are near the inside edge of the fire-line must be extinguished. Fire can cross a fire-line beneath the ground surface by burning along root channels. Therefore, burned stumps and roots near the fire-line may have to be excavated and soaked with water. Successful mopping-up makes the fire-line safe. Fire in the interior of the burned area may not be officially declared extinguished until much later.

Prescribed Burning

Prescribed burning is practiced to reduce buildup of fuels; improve grazing conditions; create seedbeds for natural regeneration of tree and shrub species and seeding of forage species; improve soil fertility through accumulation of ash; or improve wildlife habitat conditions. Therefore, prescribed burning is a management tool that can benefit the functioning of a natural ecosystem if these fires are prescribed carefully and satisfy the stated purpose.

Prescribed burning requires careful planning to minimize risks and enhance the likelihood that the objectives of the burn will be accomplished. A prescribed burn should be confined to a predetermined area and the intensity of heat and rate of spread should be controlled to a level that will achieve the desired objective. A natural resources manager must work closely with a fire management specialist to prepare an appropriate *fire prescription* (Box 10.1).

A fire prescription must include estimates of the *fire intensity* and the *rate of spread* needed to accomplish the objectives of the prescribed burning. Chances for adequate smoke dispersal should also be considered. The National Fire-Danger Rating System is often helpful in identifying the days suitable for initiating prescribed burning.

While some of the objectives of prescribed burning may be met by other means (for example, seedbed preparation can also be accomplished by mechanical preparation (see Chapter 4)), prescribed burning can be the most economical option available to a natural resources manager because the use of petrochemical energy is less costly. Disadvantages of prescribed burning include the risks involved, the problem of air pollution from smoke, and in some regions there are a limited number of days suitable for burning.

Box 10.1

A Fire Prescription: An Example (Davis et al. 1969)

The following burn conditions were selected for a prescribed fire set in a ponderosa pine forest in Arizona, where the objective was consumption of three-fourths of a two- to three-inch layer of litter and duff to enhance the potential for water yield improvement, thin understory trees, and increase herbage production:

Burning Conditions	*Prescription*
Fuel moisture:	
Upper L and F layers	Six to 12 percent
Lower H layer	15 percent or more
Fuel Temperature	
Upper 1-inch	80°F, average
Air temperature	75°F, average
Wind velocity	Two to five miles per hour
Weather	Clear
Ignition pattern	Strips set into the wind or downhill and spaced ten to 20 feet

The burn consumed over 70 percent of the litter and duff. Therefore, the objective of the prescribed fire was accomplished. Other effects of the fire included a 35 percent reduction in the density of saplings and small poles and a small increase in the production of herbaceous plants. There was also a temporary reduction of fire hazard. Twenty-four years after the prescribed burn, the litter and duff layers had returned to about two-thirds of their pre-fire depth. There had been little overall change in the density of ponderosa pine trees because trees initially damaged by the fire had died and the loss was offset by the increased growth of residual trees. The level of herbage production decreased in response to the return of litter and duff layers to pre-fire conditions (Ffolliott and Guertin 1990). Based on this example, it appears that many of the benefits from prescribed burning will be temporary unless fire is scheduled at regular intervals.

Prescribed Natural Fire

A *prescribed natural fire* is allowed to burn as long as it meets prescribed conditions. When a natural ignition is evaluated to determine the actions to take, such as allowing the fire to burn, there are risks that a fire specialist must consider, including the fire's intensity, direction, rate of spread, and severity. Natural resources managers and fire specialists must recognize and quantify these risks before decision makers are likely to accept a recommendation of allowing a natural ignition to burn.

INSECT PESTS AND DISEASES

Most insect pests and diseases are selective in their choice of a host. However, trees, shrubs, and herbaceous plants in their natural environments normally attain a state of equilibrium with indigenous pests. Outbreaks of insects and diseases occur when this equilibrium is upset. Furthermore, exotic pests are often introduced to an ecosystem when exotic trees, shrubs, or other herbaceous species are planted in an area. These pests frequently adapt themselves to the conditions of their new habitat.

Risk of damages from insects and diseases are highest when their hosts are physiologically weak. Among the causes for weakening of host plants are: planting on unsuitable sites, improper site preparation, inefficient planting techniques, adverse climatic conditions, or when maintenance operations, such as thinning or weeding, are neglected. But even healthy plants can be attacked by insects or infected with disease.

Classification

Insect pests and diseases are grouped into classes based on the way they attack the host and their general habitat. Insect pests are grouped into the following classes:

- Defoliators—Insects that eat leaves and tree needles, for example, caterpillars, cankerworms, tussock moth, sawflies, leaf beetles, and leafminers.
- Bark beetles—Insects that bore through the bark of hosts to al-

low females to lay eggs. The larvae then feed on the phloem and xylem, girdle the stem, interrupt the translocation of nutrients and moisture, and ultimately kill the host. This class includes beetles that act as carriers of lethal pathogens (the pine bark beetle) and beetles that are vectors for pathogens (beetles causing Dutch elm disease).

- Sucking insects—Insects with mouthparts that enable them to pierce foliage, small twigs, or roots to suck sap or resin for food, for example, aphids, scales, and spittlebugs.
- Wood borers—Beetles, moths, and wasps that attack hosts that are dying or otherwise suffering from stress, eventually constructing damaging tunnels to feed internally on wood.
- Shoot and bud insects—Insects that infest the meristem of their host, stunting its growth but rarely killing the plant, such as tip moths and weevils.
- Root weevils and grubs—Insects that infest the roots and root collar of their host, stunting the growth or killing the plant, the latter are the larvae of May and June beetles.

Diseases are grouped into the following classes:

- Viruses—Microscopic organisms only visible with use of an electron microscope. They cause many diseases in agricultural crops; a few viruses attack trees, for example, elm mosaic and black locust broom.
- Mycoplasms—The smallest free-living organisms; only a few cause diseases in trees, such as phloem necrosis of elm trees.
- Bacteria—Microscopic plants of single-celled bodies causing only a few diseases in trees including wetwood (a water-soaked condition of heartwood), crown gall, and discoloration and decay of the wood of some species.
- Nematodes—Microscopic eel-worms that attack roots of trees and cause dieback and abnormalities of the root systems, stunting of growth, and in some instances mortality. Examples include the pine wood nematode and other nematodes affecting trees in plantations established on abandoned farmlands.
- Mistletoes—True seed plants that are perennial evergreens parasitic on stems and the branches of trees and shrubs in mostly warmer climates, including numerous dwarf mistletoe species.
- Fungi—Plants without chlorophyll and with a simple, microscopic structure undifferentiated in stems, leaves, or roots.

Classes of fungal diseases include:

- Heartrot fungi cause decay in the heartwood of dead wood, although some fungi decay living sapwood and heartwood. Brown and white rots (sapwood fungi) decompose wood by primarily utilizing carbohydrates (cellulose) of plant cell walls, while the living heartwood fungi utilize both carbohydrates and the lignin contents of cell walls.
- Root rots cause a decline in the vigor and growth of trees by starving or destroying wood through a parasitic attack and decay of root systems, for example, the honey mushroom fungi.
- Canker diseases cause death of localized areas of bark and cambium tissues on stems, branches, and twigs of trees, for example, cankers on aspen, maples, and conifers, and one of the most notorious of all forest diseases—the chestnut blight.
- Vascular wilt diseases, including oak wilt and Dutch elm disease, cause a reduction or inhibition of normal water conduction in the xylem vessels of trees.
- Rust fungi cause infections in cones of certain confers, leaves and needles of hardwoods and conifers, and stems of conifers. These fungi require alternate hosts to complete their life cycle, examples include white pine blister rust, fusiform rust, and pine-oak rust.
- Foliage diseases cause defoliation or impairment of normal foliage functioning, resulting in a reduction in vigor and growth of some trees, although often with a relatively minor impact, examples include needle-cast fungi, brown spot needle blight, and leaf blister.
- There are multiple insect-fungus complexes that cause plant diseases, for example, the associations of blue-strain fungi with bark beetles, insects with the incidence of some vascular wilt diseases, and wood wasps with their fungal symbionts.

Control Practices

A decision to control insect pests or diseases should be based on a comprehensive review of all the factors important to the management of the impacted natural resources. The ecological, sociological, and economic implications of pest control must be weighed against anticipated benefits. When a decision to control pests is made, the tools available for this task can include chemical controls, silvicultural controls, and biological and mechanical control measures.

Chemical Control

Insect pests and diseases are often controlled by application of chemical pesticides (Box 10.2). Chemical pesticides are available as liquids or wetable powder, dusts, or smoke. Spraying with hand-operated spray guns or portable mist-blowers is frequently used to control the attack of insects and diseases on the early growth stages of plants, whether in forests, woodlands, or plantations, and on recently reseeded rangelands. When the canopy is thick, aerial spraying, dusting, or smoking can be more effective and cheaper.

Applications of chemical insecticides or fungicides into natural ecosystems can mean that some foreign elements have also been introduced. Therefore, to ensure the maintenance of environmental quality the consequences of using these control measures must be known beforehand. When these consequences suggest a detrimental effect, alternative control practices should be found.

Box 10.2

Use of Chemical Pesticides to Control Insect Pests or Diseases: Some Considerations

The use of chemical pesticides (insecticides and fungicides) to control insects or diseases requires careful planning and application. Chemical pesticides can be more damaging to the environment and the functioning of the ecosystem than the pests being controlled. Therefore, only properly tested, government-approved, environmentally sound, and locally accepted pesticides should be prescribed for use. When large-scale epidemics occur in commercial forests, there are often few practical options for control other than removing the killed, infested, or diseased trees and, if possible, utilizing them for fuel or other wood products.

Silvicultural Control

Control of insect pests and diseases is sometimes possible by thinning high-risk trees. Poor, slow growing and suppressed trees can be removed by thinning, while maintaining a vigorously growing forest stand that is able to withstand attacks by insects and diseases. The prompt removal and destruction of insect-infested or diseased trees and shrubs can also be effective in preventing the spread of pest attacks to other ecosystems.

Planting a mixture of trees and other plant species to increase the level of biological diversity can also be a silvicultural control measure. One disadvantage of planting a mixture of species is that subsequent management can become complicated. This complication can be avoided by planting different tree or shrub species in alternate blocks or wide belts. This can create a barrier to the spread of an insect pest or disease from its initial point of infection.

Biological and Mechanical Control

Biological and mechanical control of insect pests has been used with some success. Biological control involves introduction of a species-specific parasite to control the insects. Mechanical control, by physically removing and destroying the pests or eliminating alternative hosts, can be effective, although costs can be high.

INTEGRATED PEST MANAGEMENT

An effective way to control pests in many situations can be through a combination of chemical, biological, cultural, and mechanical control practices. This approach to pest control is called *integrated pest management.* Implementing integrated pest management (IPM) practices have the following advantages:

- Adverse impacts from the continuous use of chemical pesticides can be prevented.
- Development of resistance to particular pesticides by pest species can be eliminated or at least reduced in some instances.
- A backup pest control system is provided in the event that any one control method fails.

IPM is a pest control strategy that is based on knowledge of the functioning, diversity, and resiliency of the ecosystem to be protected. Methodologies from several disciplines are combined into a plan that is designed to be effective, practical, economical, and protective of the environment and people's health (Box 10.3).

Box 10.3

Features of the IMP Approach to the Control of Pests

- The focus is placed on the entire pest population and their natural enemies operating within an ecosystem. The ecosystem is the *management unit* in the case of IPM.
- The objective of IPM is maintaining pest levels below a preestablished threshold. The goal of IPM is to manage rather than eradicate a pest population.
- Control measures are selected to supplement the effects of natural controls, such as parasites, predators, and adverse weather conditions.
- Alleviation of pest problems is planned as long term and regional rather than temporary and localized.
- Possible harmful side effects to the environment are minimized through IPM practices.

Pest populations and the environmental factors that influence the abundance of pests need regular monitoring and evaluation to determine when to apply IPM actions. How the monitoring activity is conducted depends on the host plant species, pest species, climate, available managerial skills, and financial resources. Relatively simple monitoring techniques and procedures, which involve little or no special equipment, have been developed for this purpose.

Large-scale applications of IPM practices are still in their infancy. However, existing knowledge and creative research are likely to result in pest management systems that integrate two seemingly incompatible conditions: the economic need to control pest damage and the ecological necessity of conserving the inherent vitality of natural ecosystems.

ECOSYSTEM HEALTH

Fire, insect pests, diseases, and people can disrupt the functioning of natural ecosystems by slowing plant growth or killing or injuring plants and other living components of an ecosystem. Natural disturbances are often essential to the functioning, diversity, and resiliency of ecosystems. Without disturbance, many ecosystems could not be self-sustaining. But, an ecosystem is considered unhealthy when the disturbance causes continuous, severe, or widespread effects that people consider unacceptable.

Ecosystem Health Defined

People have varying views on what constitutes a healthy ecosystem. A definition of *forest health* offered by the USDA Forest Service illustrates the salient points that should be included in any definition of ecosystem health. To paraphrase this definition, *ecosystem health* is a condition where the ecosystem has the capacity across the landscape for renewal; recovery from continuous, severe, or widespread disturbances; and the retention of its ecological resiliency, while meeting the current and future needs of people for desired levels of values, products, uses, and services. However, this definition reflects a compromise among the varying interests of society; therefore, it is unlikely to satisfy everyone.

As demands on ecosystems change through time, people's views of ecosystem health will change. However, two points are included in most considerations of ecosystem health:

- A healthy ecosystem maintains its function, diversity, and resiliency.
- A healthy ecosystem provides for people's needs and desires.

Status of Ecosystem Health

The status of ecosystem health is determined by:

- Monitoring the condition of the components of the ecosystem, such as plant growth, vigor, mortality, and the occurrence of lichen communities.
- Evaluating the information obtained from monitoring in the context of the accepted definition of ecosystem health.

Monitoring and evaluation are likely to show temporal changes in ecosystem health. Natural ecosystems are dynamic so some yearly change is expected. However, unexpected and large changes can be a cause for concern and lead to a further investigation into the possible causes of the changes. By knowing how and when ecosystem health changes and what the consequences of these changes are likely to be, natural resources managers can plan remedial solutions to the problems of unhealthy and unproductive ecosystems.

SUMMARY

Fire, insect pests, and diseases are always potential threats to healthy natural ecosystems. Therefore, control measures must be taken by natural resources managers to protect these ecosystems from these and other damaging agents. Integrated pest management (IPM) is often effective in providing the most comprehensive protection.

After reading this chapter, you should have some insight into how fire and pests can be controlled to maintain healthy ecosystem conditions. You should be able to:

- Explain how the threat of fire can be prevented, or when a fire occurs, explain how the fire can be detected and suppressed.
- Answer the following questions:
  - Why the USDA Forest Service's National Fire-Danger Rating System is helpful in identifying the days when the danger of fire is greatest?
  - What information is needed to prescribe a burn for meeting a management objective?
  - How can pesticides be applied to control insect pests and diseases? Why is it important to know the effects of pesticides on natural ecosystems?
- Discuss the possible advantages of IPM in comparison to a single control method and the important features of the IPM approach.
- Appreciate the concept of ecosystem health and the relationships between the occurrence of fire, pest insect and disease outbreaks, and ecosystem health.

REFERENCES

Andrews, P.L., and L. S. Bradshaw. 1992. Use of meteorological information for fire management in the United States. In: Workshop on meteorological information for forest fire management in the western Mediterranean region. Rabat, Morocco, pp. 325–332.

Bradshaw, L. S., J. E. Deeming, R. E. Burgan, and J. D. Cohen. 1983. The 1978 National Fire-Danger Rating System: Technical documentation. USDA Forest Service, General Technical Report INT-169.

Brookes, M. H., editor. 1996. Disturbance and forest health in Oregon and Washington. USDA Forest Service, General Technical Report PNW-GTR-381.

Burgan, R. E. 1988. 1988 revisions to the 1978 National Fire Danger Rating System. USDA Forest Service, Research Paper SE-273.

Coulson, R. N., and J. A. Witter. 1984. Forest entomology: Ecology and management. John Wiley & Sons, Inc., New York.

Dahms, C. W., and B. W. Geils, technical editors. 1997. An assessment of forest ecosystem health in the Southwest. USDA Forest Service, General Technical Report RM-GTR-295.

Davis, J. R., P. F. Ffolliott, and W. P. Clary. 1969. A fire prescription for consuming ponderosa pine duff. USDA Forest Service, Research Note RM-115.

DeBano, L. F., D. G. Neary, and P. F. Ffolliott. 1998. Fire effects on ecosystems. John Wiley & Sons, Inc., New York.

Demming, J. E., R. E. Burgan, and J. D. Cohen. 1979. The National Fire Danger-Rating System-1977. USDA Forest Service, General Technical Report INT-39.

Ffolliott, P. F., and D. P. Guertin. 1990. Prescribed fire in Arizona ponderosa pine forests: A 24-year case study. In: Krammes, J. S., technical coordinator. Effects of fire management of southwestern natural resources. USDA Forest Service, General Technical Report RM-191, pp. 250–254.

Fischer, W. C. 1978. Planning and evaluating prescribed fires—a standard procedure. USDA Forest Service, General Technical Report INT-43.

Liebhold, A. M., W. L. MacDonald, D. Bergdahl, and V. C. Mastro. 1995. Invasion by exotic forest pests: A threat to forest ecosystems. Forest Science Monograph 30.

Martin, R. E., H. E. Anderson, W. D. Boyer, J. H. Deterich, S. N. Hirsch, V. I. Johnson, and W. H. McNab. 1979. Effects of fire on fuels: A state-of-knowledge review. USDA Forest Service, General Technical Report WO-13.

Metcalf, R. L., and W. H. Luckmann. 1994. Introduction to insect pest management. John Wiley & Sons, Inc., New York.

Pyne, S. J., P. L. Andrews, and R. D. Laven. 1996. Introduction to wildland fire. John Wiley & Sons, Inc., New York.

Stoddard, C. H., and G. M. Stoddard. 1987. Essentials of forestry practice. John Wiley & Sons, Inc., New York.

Tainter, F. H., and F. A. Baker. 1996. Principles of forest pathology. John Wiley & Sons, Inc., New York.

Thomas, J. W., and S. Hunke. 1996. The Forest Service approach to healthy ecosystems. Journal of Forestry 94(8):14–18.

Whelan, R. J. 1997. The ecology of fire. Cambridge University Press, Cambridge, Great Britain.

11

Soil Conservation Practices

SOME LEVEL of soil erosion by water or wind is an inevitable occurrence in natural ecosystems. However, the combined actions of water and wind can be devastating on sites with sparse vegetative cover. Improper land-use practices that further reduce vegetative cover (prolonged overgrazing by livestock, overharvesting of timber, etc.) tend to accelerate soil erosion, leading to long-term impairment of potential site productivity. This chapter discusses the process of soil erosion by water and wind and the methods for controlling these erosive processes by vegetative and other management practices.

DEFINITIONS

Erosion is the process by which soil is worn away from land surfaces by the actions of water, wind, or waves along shorelines. There are three general erosion processes: surface erosion, gully erosion, and soil mass movement. *Surface erosion* is the detachment and movement of soil particles and small aggregates of soil from land surfaces by water or wind. *Gully erosion* is the detachment and transportation of individual soil particles or large soil aggregates in a well-defined channel. *Soil mass movement* is the instantaneous downslope movement of soil, rock, and debris through the force of gravity.

Soil conservation is the management of the soil resource to prevent its loss through accelerated erosion. *Soil conservation practices,* the topic of this chapter, are primarily required on sites susceptible to soil loss, including sloping surfaces with shallow soils, sites with soils of low permeability, and sites where loss of vegetation is likely.

PROCESSES OF EROSION

Natural resources managers need to have some understanding of erosion processes to prevent excessive rates of erosion from occurring. Although in-depth knowledge of erosion processes and their control may not be a reasonable expectation, managers should be familiar with the general processes of erosion and know where to obtain further information about erosion.

Water Erosion

Erosion by water involves the detachment, transportation, and deposition of soil particles. *Detachment* is the dislodgment of soil particles from the soil mass by the energy imparted to soil surfaces by falling raindrops, a primary agent of soil erosion. *Transportation* is the entrainment and movement of soil particles, or *sediment,* by surface runoff (overland flow) that results from excessive rainfall. Sediment in surface runoff then travels overland from upslope sources to its place of deposition, which can be the toe, or the lowest part, of a slope, stream channel, flood plain, or reservoir. Some sediment deposition is temporary because subsequent storm events, often several years later, can reentrain the sediment and move it further downslope. Eventually a fraction of the sediment reaches the oceans.

The processes of surface erosion, gully erosion, and soil mass movement caused by the actions of water can occur singly or in combination with each other. It is often difficult to distinguish among these erosion processes because in reality there is a continuum of erosion forms. Furthermore, it is difficult to determine whether the erosion that is taking place is the result of natural geological weathering or an acceleration of erosion processes caused by improper land-use management practices. Descriptions of surface erosion, gully erosion, and soil mass movement erosion by water are found in Dunne and Leopold (1978), El-Swaify et al. (1982), Toy and Hadley (1987), Agassi (1995), Hudson (1995), Morgan (1995), and Brooks et al. (1997).

Wind Erosion

The movement of wind on a soil surface is analogous to the action of flowing water in a stream channel. Like water, wind exhibits a turbulent flow, with a net velocity in the horizontal direct; however, eddies are formed in both the upward and downward direction. Air is compressed randomly by this turbulence, producing gusts. A hot ground surface, such as bare soil, creates thermal updrafts, which increase with turbulence. Because of their high velocity and upward eddies, gusty winds are able to dislodge small soil particles, lift them upward, and carry them away much like suspended sediment in flowing water.

Wind erosion is more diffuse than water erosion and generally occurs where winds are prevalent and vegetative cover is sparse. In dryland environments, the combined action of wind and water removes soil from land surfaces. More detailed discussions of wind erosion are found in Brandle et al. (1988), Barrow (1991), Ffolliott et al. (1995), Morgan (1995), and Brooks et al. (1997).

EROSION CONTROL

Severe erosion threatens the productivity of forests, rangelands, and agricultural lands; the financial investments made in roads and trails, downstream reservoirs and irrigation systems; and other forms of cultural improvements. Eroded watersheds can be a major source of sediment or flood-level water flow, which can threaten land productivity and human-made infrastructures. Nevertheless, large-scale erosion control efforts should only be undertaken when the costs of control are less than the probable benefits.

Water Erosion

A key to controlling erosion caused by the actions of water is protecting soil surfaces against the energy that is expended in detaching, transporting, and depositing sediment from rainfall impact and overland flow.

Surface Erosion

Surface erosion is controlled by minimizing the impacts of rainfall and, on sites where surface erosion is a potential problem, preventing overland water flow and concentrated surface runoff, or a combination of these measures. Surface erosion can be minimized with appropriate vegetative management practices, management of grazing by livestock, and careful design of roads and trails.

Vegetative Management Practices. The most effective long-term methods for controlling surface erosion are based on establishing and maintaining a protective cover of vegetation (Box 11.1). In addition to reducing the impacts of raindrops and flowing water on the land surface, ET by plants reduces the quantity of water stored in the soil between rainfall events, resulting in greater soil storage capacity and less surface runoff during a rainfall event. The probability of soil erosion can be lessened by the network of plant roots, and soil structure is improved by the organic material added through natural decay of plant leaves, stems, and branches.

Box 11.1

Selecting Plant Species to Establish a Protective Cover

Individual plant species have different attributes for controlling surface erosion, different site requirements for growth, and different cultural treatment requirements. Therefore, knowledge of available plant species that are suitable for surface erosion control is crucial. Fortunately, appropriate plant species, site preparation methods, and planting techniques for local areas are often available.

Natural resources management practices that minimize the destruction of established protective vegetation cover and associated litter accumulation reduce the impact to the land surface from falling rain. In harvesting timber, for example, operations in which only individual or small groups of trees are removed should be encouraged rather than clearing large areas (see Chapter 4). Retaining residual strips of trees and other vegetation perpendicular to the slope of the land can also reduce downslope movement of sediment.

Grazing Management Practices. Grazing by livestock that removes inordinate amounts of herbage and compacts the soil can increase surface erosion. Excessive surface erosion resulting from grazing activities can be controlled by limiting the forage use, either by reducing the numbers of livestock or controlling the distribution of livestock on a rangeland (see Chapter 3). Reducing livestock numbers is difficult to achieve in many instances because producers (ranchers) are often reluctant to reduce their herd size due to the uncertainties associated with livestock production. The distribution of livestock can be altered through the placement of water developments, salt blocks, and fences. Alternative grazing systems are also employed to control the distribution of livestock.

Eroded rangelands can be rehabilitated by seeding of forage plants. Occasionally, water bars or other mechanical methods are needed to reduce surface erosion and improve the success of revegetation efforts. No one method of improving the condition of rangelands is likely to be successful unless accompanied by proper livestock management practices.

Design of Roads and Trails. Some surface erosion will result from nearly all types of roads and trails. However, excessive surface erosion is frequently encountered when roads and trails are improperly designed and constructed. No other activity involves such extensive and concentrated soil disturbance. Therefore, road or trail networks should be designed and constructed to:

- Minimize the time that loose soil is left exposed and the size of the area disturbed.
- Prevent any eroded sediment from reaching adjacent stream channels and gullies.
- Avoid steep gradients because these sites are less stable than gradual slopes, exhibit excessive soil exposure to erosive actions of water from cuts and fills, and concentrate water more quickly than on less steep slopes.

Many surface erosion problems from roads and trails can be eliminated in the planning stage, before construction is initiated.

Mechanical Methods. Implementing mechanical methods is often the first step in controlling surface erosion. These methods are used until a protective vegetative cover becomes established. Mechanical methods are intended to promote water infiltration and reduce soil loss from runoff by retaining water on-site. Effective methods include contour furrows, contour trenches, fallow strips, pitting, and basins. The method ultimately selected depends on the surface runoff potential of the site. A combination of mechanical methods and vegetative management practices is often prescribed. However, the establishment of protective vegetative cover should quickly follow the use of a mechanical method, regardless of the method selected.

Gully Erosion

Gully control must stabilize the slope gradient and nickpoints (Box 11.2) of be effective. Permanent gully control is obtained only by returning the eroded landscape to a *watershed good condition* (see Chapter 2). Achieving this condition necessitates the establishment and maintenance of a protective vegetation cover not only on severely eroding sites but also on sites above the area where surface runoff originates.

Vegetative Management Practices. Erosion on slope gradients can be controlled by establishing a rapidly growing vegetative cover, often without the need to use structural measures. The most effective vegetative cover for these sites is comprised of high density, fast-growing, relatively short herbaceous plants with deep and dense root systems. Tall, flexible herbaceous plants are less suitable. Flowing water forces these plants to the bottom of the gully, creating a smooth path for water to flow to the stream and ultimately increases runoff velocity. Trees and shrubs are less effective in controlling gully erosion because they can restrict the flow of water and cause a diversion of water against streambanks. Restrictions to flowing water in the original gully can result in the development of new gullies or head cuts when the water reenters the original gully.

Trees and shrubs are often planted in wide gullies on low sloping gradients to form *live dams,* which reduces runoff velocity and allows sediment to accumulate. But, establishing vegetation on actively eroding sites can take a considerable period of time. Therefore, structural measures (see below) may be required to temporarily stabilize a site until vegetation can become established. During this time, it is necessary to exclude livestock from the eroded site and follow-up with strict grazing controls. Structural measures alone should not be considered a permanent solution to control gully erosion regardless of how well they are constructed.

Box 11.2

Nickpoints: An Explanation

Nickpoints occur in gully channels at locations where there is an abrupt change of elevation and slope and a lack of protective cover. Water falling over a nickpoint causes the soil underlying the nickpoint to erode, the nickpoint then migrates uphill (head cutting). The force of falling water simultaneously dislodges sediment below the fall and transports it downhill, lengthening and deepening the gully in the downhill direction (down cutting).

Structural Measures. Structural measures become necessary if conditions do not permit the immediate establishment of a vegetative cover. Structural measures can be required at critical locations where stream channel changes occur. Critical locations are places where deepening, widening, and deposition frequently alternate with different regimes of water flow, for example, nickpoints, head cuts, and gully reaches close to the gully mouth. Normally, these critical locations can be easily identified.

An effective structural design will help to establish and maintain vegetation. Vegetation can become established on the gully bottom if the gully gradient is stabilized. Stabilized gully bottoms help stabilize streambanks, since the toe of the gully side slopes is not likely to move. This process is often mechanically accelerated by sloughing gully banks, particularly where steep banks prevent vegetation establishment. However, gully banks should be sloughed only after the bottom of the gully is stable. Vegetative establishment is also accelerated if large deposits of sediment are allowed to accumulate in the gully above structural works. These sediment deposits increase soil-water storage capacities and decrease channel gradients.

One structural measure that is frequently used in attempting to control gully erosion is the construction of *check dams*. The purpose of a check dam is to replace the V-pattern of channel bed erosion with a broader channel bottom, which is formed by the retention of sediment and other debris behind the dam. This reduces the tractive forces of the erosion processes. Check dams can also support the toe of steep slopes by sediment deposition and protect slopes from undercutting. Sediment that is deposited behind a check dam:

- Develops a new channel bottom with a gentler slope than the original gully bottom, reducing the velocity and erosive forces of gully flows.
- Stabilizes side slopes of the gully and allows slopes to adjust to their natural angle, reducing any further erosion of channel banks.
- Promotes the establishment of vegetation on the slopes and channel bottom of a gully.
- Stores and slowly releases water, raising the local water table and enhancing vegetative growth outside the gully.

Check dams can be built of loose rock, rock bound by wire mesh, or nonporous materials, such as prestressed concrete. Logs and small branches have also been used in combination with loose rocks in the construction of check dams. However, these materials do not always remain in place long enough to sustain the control of gully erosion. Rotting wood can cause changes in the conformation of the check dam, which may result in a new cycle of erosion.

When a series of check dams are constructed, the first check dam should be located in the gully where down cutting does not occur; that is, it should be located where sediment has been deposited at the mouth of the gully, or where there is a rock outcrop or maintained road crossing, or where the gully enters a stream system. Spacing of subsequent dams upstream from the first dam depends on the gradient of the gully floor, the gradient of the "sedi-

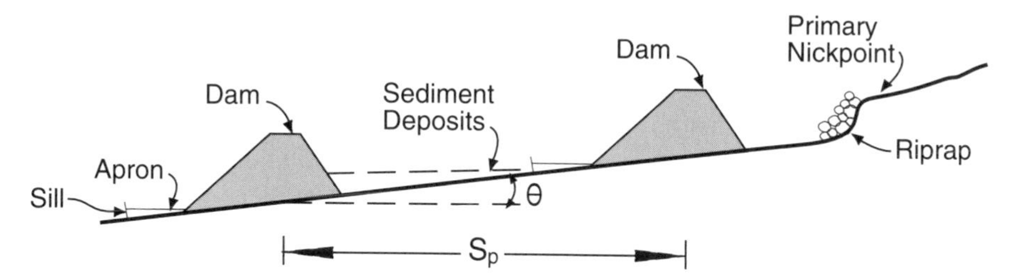

Figure 11.1. Placement of a check dam. [S_p = spacing, θ = angle of gully gradient] (from Brooks et al. 1997).

[A calculation for the spacing of check dams (S_p) developed by Heede and Mulfich (1973) is based on the effective height of the dam from the gully bottom to the spillway crest, the angle corresponding to the gully gradient (θ), and a constant related to the gradient of the sediment deposits.]

ment wedges" deposited upstream of the dams, and the effective height of the dams, which is measured from the gully floor to the bottom of the spillway. A diagram of the placement of check dams is presented in Figure 11.1.

Soil Mass Movement

Deep-penetrating and dense root systems of trees and large shrubs contribute to the fractional resistance of a sloping soil mass.

Trees and shrubs help stabilize soils by vertically anchoring themselves into stable substrate. Plants with fine to medium root systems provide lateral strength and also improve slope stability. Removal of soil water by ET on a hillslope can result in lower water pore pressure, reduced weight of a soil mass, and reduced chemical weathering. The root strength characteristics and high ET rates of tree and shrub species suggest that a forest, woodland, or plantation cover is better in reducing soil mass movement than is a cover of herbaceous plants. Therefore, properly planned and carefully implemented forestry practices are a key to the control of soil mass movement.

Road construction and tree harvesting have major land-use impacts that can lead to soil mass movement. Timber harvesting practices on steep slopes, such as the removal of trees from a hillslope, must be done in a manner that will maintain stability of the soil mass. Undercutting of slopes and improper drainage of roads are also major causes of accelerated soil mass movement. Sites that are susceptible to soil mass movement should be avoided when planning road construction and timber harvesting. Similar considerations must be taken into account when trees and shrubs are removed in the conversion of a forest or woodland to an agricultural land.

Wind Erosion

Erosion caused by the actions of wind is best controlled by reducing the wind velocity near the soil surface. Wind velocities are reduced and the magnitude of wind erosion decreased by planting trees, shrubs, or occasionally herbaceous plants to form a windbreak (Figure 5.1). Windbreaks are barriers established to protect a site from wind flows by reducing its velocity. They can be composed of woody or nonwoody plants or nonvegetative material (board or rock walls, earthen banks, etc.). The focus of this section is placed on use of trees and shrubs in the creation of windbreaks.

Selection of Tree and Shrub Species

Tree and shrub species used in windbreaks must be adapted to the soil types and climate of the area. The trees and shrubs selected for windbreak plantings should possess the following characteristics:

- Resistance to the force of wind.
- Rapid vertical growth.
- Straight stems, wind firmness, dense and uniform crown formation.
- Deep root systems that do not spread into agricultural fields or pasture lands. Laterally rooted species can compete with the agricultural crops and forage resources they are suppose to protect.
- Resistance to drought, diseases, insects, and other pests.
- Appropriate phenological foliage characteristics, for example, year-round leaves or leaves present only part of the year.

Rarely will one species have all the above characteristics. Therefore, a combination of two or more tree or shrub species is frequently needed to provide an adequate barrier to the erosive effects of wind.

Other Design Criteria

Three zones are delineated in a windbreak planting: a *windward zone* which faces the direction from which the wind blows, a *leeward zone* which is the side of a windbreak that is sheltered from the wind, and a *protected zone* where the effects of windbreak planting are felt. The relative widths of these zones are dependent on:

- The orientation and pattern of the windbreaks with respect to the prevailing winds.
- The height of the barrier.
- Its density.
- The spacing within and between windbreaks and their internal structure.

Other criteria for the establishment of windbreaks are found in Brandle et al. (1988), Barrow (1991), Ffolliott et al. (1995), and Morgan (1995).

Management Considerations

Gaps in windbreak plantings cannot be tolerated. Replacement plantings should be considered when trees or shrubs are lost, especially in those situations where the elapsed time from the original planting is relatively short and replanting will allow the integrity of the windbreak to be retained.

Protection from wind erosion and the other benefits offered by windbreaks are often combined with the production of wood products by selecting tree and shrub species appropriate for both objectives. In doing so, the protective and productive functions of these barriers can be integrated into sustainable land management strategies.

When water resources are limited, the beneficial effects of windbreak plantings must be weighed carefully because windbreaks must be watered to survive. Adverse environmental effects can also occur if trees and shrubs comprising the windbreaks harbor birds, insects, or diseases that are harmful to agricultural crops. The possibility of a *symbiotic relationship* (two dissimilar organisms living together in close association) forming between an agricultural crop pest and windbreak plants must be thoroughly investigated before selecting the tree or shrub species to use for windbreak plantings.

PREVENTION OF SOIL LOSS

Preventing soil loss by avoiding the conditions that cause water or wind erosion or minimizing their occurrence is the most effective means of combating soil erosion (Box 11.3). However, preventative measures become necessary on sites where excessive soil loss occurs. Maintenance of protective vegetative cover is the best method for preventing soil loss. Measures should be taken to maintain a soil surface that can readily absorb water, thus preventing soil erosion. Also, when more water infiltrates into the soil, the better the chance of sustaining plant growth and a protective vegetative cover. Preventative measures include:

- Avoiding land management practices that reduce water infiltration capacity and soil permeability.
- Maintaining a vegetative cover for as long as possible each growing season.
- Retaining vegetative cover on hilltops or steep slopes and restricting or eliminating livestock grazing on these areas.
- Locating livestock watering facilities so minimal runoff from the facilities reaches downstream water bodies.
- Carefully designing and constructing roads and trails, especially on sites where water flow is likely to become concentrated.

Box 11.3

The Critical Point of Deterioration: A Principle (Satterlund and Adams 1992)

The importance of using preventative measures in soil conservation is illustrated by the *critical point of deterioration* principle. This principle states that for every site there is a point of erosion (soil loss) beyond which further deterioration occurs rapidly and natural stabilizing forces are incapable of overcoming. In general, the principle applies because accelerated erosion deteriorates a site at an ever-increasing rate unless the processes are reversed. A representation of the principle on a moderately stable site is shown below.

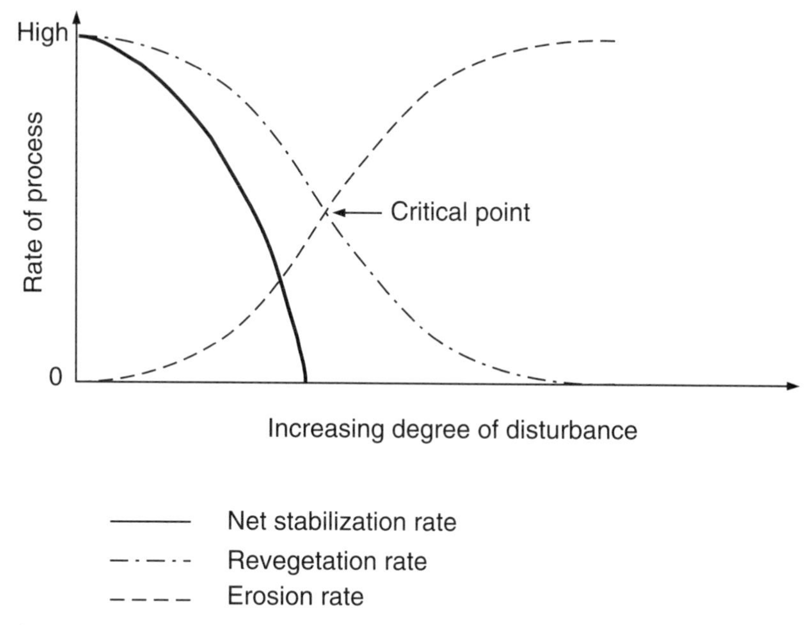

Accelerated erosion becomes self-sustaining at the *critical point of deterioration* and the site cannot recover by the natural process of revegetation. To the right of the critical point, deterioration continues to completion. To the left, a site recovers its stability at an increasing rate (from Satterlund and Adams 1992).

SUMMARY

Soil erosion by water and wind is a natural phenomenon. However, overgrazing by livestock, overharvesting of timber, or other improper land-use management practices often accelerate the rate of erosion, especially on sites where these management practices have reduced the protective vegetative cover.

After reading this chapter, you should have an understanding of the processes of soil erosion, the importance of soil conservation practices, and be able to:

- Describe the general processes of soil erosion by the actions of water and wind.
- Explain how natural resources management practices and changes in vegetative cover influence the processes of soil erosion.

- Discuss the following with respect to the actions of water:
  - The processes of surface erosion, gully erosion, and soil mass movement on upslope sites.
  - Actions that can be taken to control soil erosion, including vegetative management practices, livestock grazing management practices, and mechanical methods of control.
- Discuss the following with respect to the actions of wind:
  - How wind moves soil.
  - The importance of selecting suitable trees and shrubs and other design criteria for planning windbreak plantings.
- Understand the role of soil conservation practices in preventing soil erosion.

REFERENCES

Agassi, M., editor. 1995. Soil erosion, conservation, and rehabilitation. Marcel Decker, New York.

Barrow, C. J. 1991. Land degradation. Cambridge University Press, Cambridge, Great Britain.

Brandle, J. R., D. L. Hintz, and J. W. Sturrock, editors. 1988. Windbreak technology. Elsevier, New York.

Brooks, K. N., P. F. Ffolliott, H. M. Gregersen, and L. F. DeBano. 1997. Hydrology and the management of watersheds. Iowa State University Press, Ames, Iowa.

Dunne, T., and L. B. Leopold. 1978. Water in environmental planning. W. H. Freeman and Company, San Francisco.

El-Swaify, S. A., E. W. Dangler, and C. C. Armstrong. 1982. Soil erosion by water in the tropics. Research Extension Service 024, College of Tropical Agriculture and Human Resources, University of Hawaii, Honolulu, Hawaii.

Ffolliott, P. F., K. N. Brooks, H. M. Gregersen, and A. L. Lundgren. 1995. Dryland forestry: Planning and management. John Wiley & Sons, Inc., New York.

Heede, B. H. 1976. Gully development and control: The status of our knowledge. USDA Forest Service, Research Paper RM-169.

Heede, R. H., and J. G. Mulfich. 1973. Functional relationships and a computer program for structural gully control. Journal of Environmental Management 1:321–344.

Hudson, N. W. 1995. Soil conservation. Iowa State University Press, Ames, Iowa.

Lal, R., and F. J. Pierce, editors. 1991. Soil management for sustainability. Soil and Water Conservation Society, Ankeny, Iowa.
Morgan, R. P. C. 1995. Soil erosion and conservation. Longman Group Limited, Essex, England.
Satterlund, D. R., and P. W. Adams. 1992. Wildland watershed management. John Wiley & Sons, Inc., New York.
Tinus, R. W., editor. 1976. Shelterbelts on the Great Plains: Proceedings of the symposium. Great Plains Agricultural Council Publication 78.
Toy, T. J., and R. F. Hadley. 1987. Geomorphology and reclamation of disturbed lands. Academic Press, Orlando, Florida.
Troeh, F. R., J. A. Hobbs, and R. L. Donahue. 1991. Soil and water conservation. Prentice-Hall, Englewood Cliffs, New Jersey.

12

Rehabilitation of Disturbed Lands

OTHERWISE PRODUCTIVE lands can be laid bare because of a destructive wildfire, overgrazing by livestock, overharvesting of timber, or degrading mining operations. These disturbances often cause increased soil erosion and a loss in site productivity. Regardless of the cause, disturbed lands are often at risk until their ecological functioning is restored through some form of rehabilitation, which is the subject of this chapter.

DEFINITIONS

Disturbed lands are lands that have undergone or are undergoing degradation or destruction by excessive soil erosion, salinization, or combinations of these processes. *Rehabilitation* is the act of returning disturbed lands to a more useful state. The term *restoration* is often used as synonym of rehabilitation; however, restoration refers to returning lands to their *original state,* which is often unknown or if known is impossible to attain.

Rehabilitation methods are applications of natural resources management practices and mechanical or structural measures to reverse the processes of degradation and destruction on otherwise productive lands. Rehabilitation methods are based on proper application of the natural resources management practices discussed in this book.

REHABILITATION METHODS

Establishing a protective vegetative cover is the most effective method to successfully rehabilitate disturbed lands. Vegetative cover can control soil loss and the loss of site productivity by minimizing the impacts of falling raindrops, flowing surface water, and wind blowing across the soil surface (see Chapter 11). However, contour furrows, terracing, or the construction of check dams may also be necessary to retain water on-site until a vegetative cover can become established. These mechanical methods promote infiltration and reduce soil loss from runoff.

The first task in establishing a protective vegetative cover is often substrate improvement. The substrate exposed by a wildfire or from improper natural resources management practices can be too acid or too alkaline, or it can be infertile or contain toxic substances, exhibit poor infiltration or drainage properties, and have poor physical structure or contain impermeable crusts. Topsoil is often the best substrate to use. Other substrates can be used, but structure and fertility improvements may be necessary. For example, when seeding and planting are not feasible, mulch can be applied to protect the land surface from soil loss. Other forms of site preparation can also be required to ensure successful revegetation.

Selection of plant species to revegetate disturbed lands depends on the goals of the rehabilitation, the condition of the disturbed site, and the availability of seed or planting stock. Rehabilitation goals include avoiding increased soil erosion, preventing stream pollution, or restoring a lost plant community. It is usually impossible to restore a *climax plant community* because only plant species representative of earlier stages of succession are likely to be adapted to severely disturbed sites.

Sources of information on suitable planting stock to revegetate disturbed sites are found in Thames (1977), Cox et al. (1984), Oechel (1988), Roundy et al. (1995), and other literature on the rehabilitation of disturbed lands.

Following Destructive Wildfire

Wildfire is a natural part of many ecosystems. However, wildfires can drastically disturb a landscape by increasing soil erosion rates, changing plant species composition from herbaceous to unpalatable nonforage plants, and causing excessive losses in timber resources (see Chapter 10). The severity of post-fire degradation is largely dependent on:

• The intensity of the fire.
• The size of the burned area.
• The characteristics of the burned site.
• The rate of ecosystem recovery from fire effects.

There are situations where wildfire effects are negligible, for example, where a low-intensity surface fire only partially removes organic layers, or where there is quick regrowth after a fire on a dormant grassland, or a fire occurs on a level terrain with high infiltration rates. Even in situations where a wildfire has consumed all of the organic material, the post-fire site conditions may include only minor erosive problems and damage to water quality, thus permitting a rapid recovery. Rehabilitation is usually not necessary on these sites.

However, a major fire can cause extensive post-fire soil erosion. When intense rainstorms strike the soil surface on burned sites before a protective vegetative cover has been reestablished, severe erosion can result. Post-fire flooding and other devastating impacts can also occur. In these cases, rehabilitation measures, which usually involve the establishment of a protective vegetative cover, are often necessary to accelerate post-fire recovery. Emergency revegetation efforts may be initiated to establish a temporary plant cover immediately following a wildfire (Box 12.1). Ideally, this is a transitory vegetation cover that is then replaced by more permanent plant species.

Fire suppression activities can also cause land disturbances. For example, improperly constructed or aligned fire-lines can increase soil erosion and limit the success of establishing a protective plant cover after a fire. Therefore, fire-lines should be constructed carefully (for example, across contour lines) to control post-fire soil erosion (see Chapter 10). Closely spaced water bars within the fire-lines often remedy this problem.

Overgrazed Rangelands

Impacts on rangelands by grazing livestock are minimal when proper rangeland management practices are followed (see Chapter

3). However, problems can occur and rehabilitation can become necessary when prolonged overgrazing increases soil erosion and losses in forage productivity. These rehabilitation measures include:

- Reducing the stocking rates of livestock, improving the distribution of livestock on the rangeland, or combinations of both, and through modification of grazing practices.
- Initiating needed rangeland improvement techniques, such as developing additional water, removing undesirable plants to favor the production of forage, or seeding of forage species.

Box 12.1

Emergency Revegetation Following Wildfire: An Exception

Emergency revegetation is not always the proper action to take following a wildfire. To illustrate this point, emergency seeding of perennial grass species on frequently burned chaparral watersheds of southern California is a common practice (Keeley and Keeley 1981, Conrad and Oechel 1982). On some sites, however, this seeding can delay the post-fire recovery of sclerophyllous shrub species, has little potential for reducing sediment yields, and can increase erosion rates and soil loss.

Few rangeland improvements are accomplished if livestock numbers are high and grazing intensities are excessive. In these situations, attempts should be made to avoid heavy grazing in localized areas by reducing livestock stocking rates and achieving a better distribution of the livestock. Elimination of livestock can be the best solution to the problem if the rangeland is severely disturbed, although this action is likely to meet with resistance from the livestock producers (ranchers).

Rangeland improvements (see Chapter 3) can be initiated once the livestock numbers and distribution problems have been mitigated. Available options must be considered carefully because of the high costs of most rangeland improvement practices. The following guidelines are helpful:

- Improvement methods should be appropriate for the rangeland condition or stage of deterioration.
- Methods that have been locally successful are preferred over methods used in other regions.
- Small-scale pilot studies should be carried out before widespread implementation of methods used in other regions.
- Changes in rangeland management practices resulting from implemented improvement methods, such as modifying grazing systems to allow reseeding to become established, must be compatible with the goals of livestock producers and those involved in natural resources management.
- Improvement methods should be carried out in locations where the potential for improvement is greatest.

Overharvested Forest Lands

Timber on commercial forest lands is harvested for processing into lumber, pulp, or other wood products. Timber production can be sustained within the framework of proper stocking conditions if the allowable cut has been calculated correctly and not exceeded by the harvesting intensity. Timber harvesting operations should also comply with established environmental regulations (see Chapter 4). However, rehabilitation measures become necessary where timber harvesting has been excessive, or when timber harvesting operations result in increased rates of soil erosion, which threatens long-term forest productivity and other uses of the ecosystem.

The methods used to rehabilitate overharvested forest lands to ensure future timber production are dependent on whether regeneration is natural or artificial. When the likelihood for a forest to achieve natural regeneration is small or when the time required to achieve successful natural regeneration is long, a forester may opt for artificial regeneration. A forester's goal is to quickly place the harvested forest back into a condition of productivity.

Regardless of whether the regeneration is natural or artificial, stabilizing the disturbed land to prevent further soil losses is a necessary first task in the process of successful forest regeneration. Surface erosion, the most immediate cause of increased soil loss following a disturbance, is often initially controlled by mechanical methods. Mechanical methods allow protective plant cover to become established, if this is an objective of the rehabilitative effort (see Chapter 11). Road closures, construction of water bars to divert flowing water off of the roadways, and seeding the roadbed with herbaceous plants can be necessary to prevent concentrated flows of runoff water in the roads. Once the land has been adequately stabilized, one of the following actions is taken:

- The appropriate silvicultural system allowing the forest community (type) to reproduce is selected. This will become the foundation for future planning of timber management practices and, when regeneration is natural, other cultural treatments (see Table 4.1). This silvicultural system is selected based on:
  - The silvical characteristics (tolerance, water requirements, seeding intervals, etc.) of the tree species.
  - The receptiveness of the seedbed for germination and seedling establishment.
  - Whether even-aged or uneven-aged stand structures are the basis for future forest management.
- Artificial regeneration practices are implemented (see Chapter 4), in which case decisions are based on:
  - Whether the forest will be comprised of the tree species previously harvested or converted to another species.
  - Whether site preparation, other than the soil scarification resulting from the harvesting operations, is necessary.
  - What will be the spacing and spatial arrangements of the plantlets.
  - Whether a protective cover of herbaceous plants must first be established to prevent further soil loss before the forest regeneration is implemented.

Following Mining Operations

While the areas impacted by mining operations can be relatively small, they can cause disproportionate disturbances to a natural ecosystem by becoming point sources for soil erosion and pollutants. However, there are only a few mining operation activities that degrade a site to the point where rehabilitative efforts are necessary. These mining activities are:

- Construction of road systems.
- Segregation and stockpiling of overburden materials (materials overlying the mining operations) and soils.
- Shaping (piling) of spoils and tailings.
- Construction of dams and other impoundments.

The construction of road systems, dams, and other impoundments is normally the responsibility of civil engineers. Natural resources managers sometimes become involved to assure that the construction of these and other infrastructures are consistent with sustaining ecosystem integrity. Natural resources managers will more likely become involved with the following:

- Ensuring proper segregation and stockpiling of overburden materials and soils to their approximate original location.
- Shaping spoils and tailings to ensure mass stability.

Surface mines in the United States must be restored to their approximate original (pre-mining) contours following completion of the mining operations. But, complete restoration of the original contours is not always possible when the volume of material removed in the mining operation was large; or when spoils or tailings are piled on the land surface. In either case, the disturbed land must be well vegetated if it is to resist the forces of soil erosion. Restoration of land and water resources is a high priority in mining reclamation efforts.

Natural slopes normally have irregular and uneven surface patterns consisting of a hierarchy of minute stream channels. To maintain a safe flow of water, reclaimed slopes should have drainage patterns that are similar to the natural patterns that evolved through the actions of surface runoff (overland flow). However, accelerated and often excessive surface runoff can occur in the time interval between the shaping of the slope and the reestablishment of vegetation. Therefore, the restored stream channels into which surface runoff flows must be protected with rock or other appropriate materials. Sediment basins should also be constructed to trap eroded sediment.

Spoils and tailings must be shaped, or piled, to ensure mass stability on the selected piling surface. A convex piling surface will direct drainage water to move over, rather than through, accumulated spoils or tailings that contain toxic materials. It may also be necessary to construct a network of stream channels that collect surface runoff for rapid flow from the site, limiting the percolation and leaching of toxic materials. It is generally better to create a channel network that is denser than the natural network rather than one that is sparser.

REHABILITATION OF SALT-AFFECTED LANDS

Excessive reductions in vegetative cover resulting from improper natural resources management activities often lead to soil erosion (see Chapter 11). What is less well known, however, is that in dryland environments these reductions in vegetative cover can also lead to salinization problems. Through an understanding of the process of salinization (Box 12.2) and the impacts of revegetation on this process, it is possible to rehabilitate salt-affected lands by planting salt-tolerant plants.

Box 12.2

Causes of Salinization in Dryland Environments

High rates of ET relative to the limited percolation of infiltrated water through the soil are a common feature of dryland environments. This condition allows salt, a natural constituent of rain, to accumulate and persist in the upper layers of a soil profile. The characteristically low rainfall amounts of drylands are often insufficient to flush the salt to groundwater reserves. Therefore, salt concentrations can reach high levels in the rooting zones of plants.

Reductions in the leaf area of vegetation, which could result from overgrazing by livestock, reduces ET, with the water savings from this reduction increasing throughfall (the rain that falls through the canopy), runoff, infiltration, soil water storage, percolation of infiltrated water to groundwater aquifers, and seepage. When this happens, the water table rises to the level of salt accumulated in the upper layers of the soil, which moves the accumulated salt upward toward the soil surface. This is the process of *salinization.*

Revegetation can be used on sites with low leaf area. For example, a cover of forage crops can be planted on a rangeland where overgrazing by livestock has reduced leaf area, root growth, and transpiration rates; or trees and shrubs can be planted on sites that support only a sparse cover of herbaceous plants as a result of overgrazing. Regardless of the reasons for increasing leaf area, the salinization process that was caused by leaf area reduction are generally reversed with successful revegetation.

Revegetation

Rehabilitation of a site with saline soil through revegetation requires a careful selection of salt-tolerant plants. A primary criterion for plant selection relates to the need to reduce the recharge of groundwater resources; therefore, plants with superior evaporating surfaces should be selected. Points to consider in the plant-selection process include:

- Plants with deep roots transpire more water than plants with shallow roots; therefore, deep-rooted plants should be favored.
- Perennial plants should be favored over annual plants because annuals generally have smaller leaf area, shorter leaf-cover duration, and shallower roots.
- Trees and tall shrubs should be favored over herbaceous plants. The elevated canopies of trees and tall shrubs make better use of atmospheric advective energy, which leads to increased evaporation from wet leaves (intercepted rainfall) and higher evaporative demands (vapor pressure deficits).
- Evergreen plants transpire more water than deciduous plants because the former have longer leaf-cover duration.
- Phreatophytes transpire more water than nonphreatophytes because water supplies are less limiting. This situation can be modified if salinity in saturated soils is high, the depths to phreatical surfaces are great, or the oxygen content is low.

Trees and shrubs are established on salt-affected lands as seedlings, from cuttings or other vegetative plant sections, or through sowing of seeds. Herbaceous plants are generally established through seeding. Once a successful planting is achieved, the following guidelines should be considered:

- The possibilities of waterlogging, flooding, and soil erosion are minimized by protecting the planting sites with appropriate soil conservation measures.
- Salt levels at the surface of soils are reduced by cultivating the planting sites before initial seasonal rains to encourage infiltration water into the soils and to leach salts from topsoil.

Other important considerations include:

- Planted areas may require protection from grazing by livestock and damage from wildlife activities.
- Protective measures should be implemented where wind erosion is a problem.
- There should be early detection and subsequent control of insect pests and diseases.

Revegetation Strategies

When planning the rehabilitation of salt-affected lands by revegetation practices, the lands should be separated into recharge and discharge areas. *Recharge areas* are sites where the causes of salinization occur, while *discharge areas* are the sites where the results of salinization are apparent. Immediate revegetation of recharge areas is mandatory to reduce the recharge of groundwater reservoirs, while the treatment of discharge areas can often be deferred until later.

Revegetation is generally more successful on recharge areas where salinity levels are usually low. Revegetation is most effective if the plant roots reach into the permanent supply of water in the saturated zones of soil. When roots are near or in the water table, a maximum dispersion of plants is the most effective placement of the plantlets. Land-use practices and the manner of managing the plants often override this consideration, however.

Two processes must take place before sites with saline soil can once again support salt-sensitive vegetation:

- The water table must fall below the level where salt has accumulated in the upper soil profile.
- After the water tables have retracted to safe levels, salt must be leached from the soil by percolation for a specified period of time.

Achieving rehabilitation of saline sites requires several years, because it takes time for deep-rooted plants to attain their maximum ET capacities. On the other hand, short-rooted and less-effective annual plants reach their maximum potentials in the first year of sowing. However, even when full ET capacities are reached, the evidence of rehabilitative processes may not be evident for a number of years.

SUMMARY

Otherwise productive lands can become unproductive following a destructive wildfire and improper natural resources management practices. When these disturbances occur, rehabilitative measures are often required to restore the ecological functioning of the disturbed lands. One form of rehabilitation is the planting of trees, shrubs, or other herbaceous species after proper site preparation.

After reading this chapter, you should be able to:

- Appreciate the need to quickly revegetate severely burned sites.
- Discuss the methods that can be used to rehabilitate overgrazed rangelands and overharvested commercial forest lands.
- Explain why the segregation and stockpiling of overburden materials and soils and the stabilization of spoils and tailings are important to rehabilitation of landscapes affected by mining operations.
- Explain the process of salinization and the impacts of vegetative management on this process.

REFERENCES

Agassi, M., editor. 1995. Soil erosion, conservation, and rehabilitation. Marcel Decker, New York.

Allen, E. B., editor. 1988. The reconstruction of disturbed arid lands: An ecological approach. Westview Press, Inc., Boulder, Colorado.

Barrow, C. J. 1994. Land degradation. Cambridge University Press, Cambridge, Great Britain.

Brown, P. L., A. D. Halvorson, F. H. Siddoway, H. F. Mayland, and M. R. Miller, 1983. Saline-seep diagnosis, control and reclamation. USDA Agricultural Research Service, Conservation Research Report 30.

Cairns, J., Jr. 1991. Rehabilitating damaged ecosystems: Volume I. CRC Press, Inc., Boca Raton, Florida.

Conrad, C. E., and W. C. Oechel, technical coordinators. 1982. Dynamics and management of Mediterranean-type ecosystems. USDA Forest Service, General Technical Report PSW-58.

Cox, J. R., H. L. Morton, T. N. Johnsen, Jr., G. L. Jordan, S. C. Martin, and L. C. Fierro. 1984. Vegetation restoration in the Chihuahuan and Sonoran Deserts of North America. Rangelands 6:112–115.

Greenwood, E. A. N. 1988. The hydrologic role of vegetation in the development and reclamation of dryland salinity. In: Allen, E. B., editor. The reconstruction of disturbed arid lands. Westview Press, Inc., Boulder, Colorado, pp. 205–233.

Keeley, J. E., and S. C. Keeley. 1981. Post-fire regeneration of southern California chaparral. American Journal of Botany 68:524–530.

McKell, C. M. 1986. Propagation and establishment of plants on arid saline land. In: Barrett-Lennard, E. G, editor. Forage and fuel production from salt-affected wasteland. Elsevier, Amsterdam, The Netherlands, pp. 363–375.

Morgan, R. P. C. 1995. Soil erosion and conservation. Longman Group Limited, Essex, England.

Oechel, W. C. 1988. Seedling establishment and water relations after fire in a Mediterranean ecosystem. In: Allen, E. B., editor. The reconstruction of disturbed arid lands. Westview Press, Inc., Boulder, Colorado, pp. 34–45.

Roundy, B. A., E. D. McArthur, J. S. Haley, and D. K. Mann, compilers. 1995. Wildland shrub and arid land restoration symposium. USDA Forest Service, General Technical Report INT-315.

Sauer, L. J. 1998. The once and future forest: A guide to forest restoration strategies. Inland Press, Covelo, California.

Thames, J. L., editor. 1977. Reclamation and use of disturbed land in the Southwest. University of Arizona Press, Tucson, Arizona.

Toy, T. J., and R. F. Hadley. 1987. Geomorphology and reclamation of disturbed lands. Academic Press, Orlando, Florida.

Vallentine, J. F. 1989. Range development and improvement. Academic Press, Orlando, Florida.

13

Integrated Natural Resources Management

THE PROBLEMS with integrating natural resources management into a multiple-use, ecosystem-based concept are not always fully recognized by planners, managers, and decision makers. While problems can often be structured in terms of resource-oriented, multiple-use management objectives, people must also be aware of area-oriented, multiple-use management implications. This awareness is especially important when natural resources management plans are developed and implemented on large areas.

Natural resources management involves the development, application, and evaluation of management practices designed for one or more specified purposes, for example, to increase water, livestock, wood production, or a combination of these. However, the impacts of a management practice are likely to extend far beyond the attempt to increase water flow, livestock, or wood. People can demand other natural resource products and uses (Figure 13.1). For this reason, planning the use of a management practice that increases livestock, water, or wood production, or any natural resource product or use should be based on the broad, multiple-use, ecosystem-based concept.

IMPORTANCE OF ALTERNATIVES

Alternative plans for the use of natural resources management practices should be developed because there are usually a number

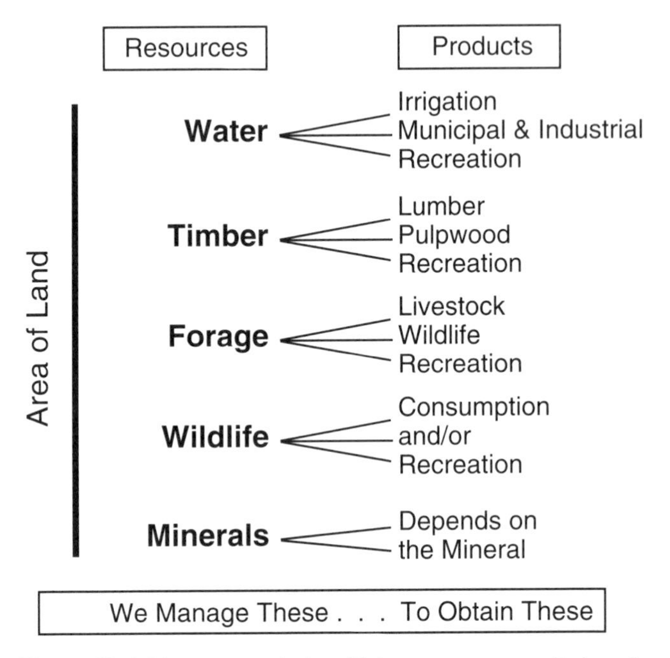

Figure 13.1. Management of multiple resources results in natural resource products and uses that are demanded by people (from Ridd 1965).

of ways to achieve a specified management goal. One alternative is to allow conditions to remain as they are; that is, what is happening at the present time could be the best natural resources management practice. Other management alternatives will involve some form of change.

In selecting the best natural resources management practice to implement, it is necessary to determine the advantages and disadvantages of each in terms of the multiple-use, ecosystem-based management potential present *before* the practice is implemented. It becomes necessary to:

- Estimate the level of natural resources production or use for all of the alternative natural resources management practices being considered.
- Determine the benefits and costs associated with implementation of each alternative.
- Conduct economic evaluations of each management practice, and

then select the best practice to implement on the basis of these economic evaluations or other decision-making procedures.

ESTIMATION OF NATURAL RESOURCES PRODUCTION AND USE

Determining the response of each alternative management practice being considered requires estimating the level of natural resources production and use through on-site measurements and inventories or computer simulation techniques (see Appendix 4). Estimates of what the levels of production or use will be *before* and *after* a natural resources management practice is implemented are needed. These estimates often include:

- Water yields and quality characteristics.
- Forage production levels for subsequent translation into livestock gains.
- Wood production and quality of wood products.
- Wildlife habitat potentials.
- Outdoor recreational values and scenic values.

The estimates of natural resources production and use can be summarized in a *product mix table*. This table presents multiple-use options by quantitatively representing all the estimated natural resource products and uses obtainable from a watershed; a livestock grazing allotment; or a forest, woodland, or plantation. Estimates for a site before a proposed management practice is implemented (that is, the "as is" situation) are used for comparison with the estimates of the conditions after implementation of the alternative management practices (Box 13.1). These comparisons are then used to estimate the gains and losses in multiple-use terms for each alternative management practice being considered.

Estimates presented in a product mix table typically represent *average responses*. It is also helpful to know the *variability* of these responses. For example, the amount of forage produced for livestock consumption in a year is generally linked to rainfall patterns; the greater the variability in rainfall, the greater the variability in the estimated average forage production. A measure of this variability is necessary to determine any statistically significant differences in the estimated forage production among the alternative management practices (see Appendix 3). Similar measures of variability are needed for the other natural resource products and uses being considered.

Box 13.1

Estimates of Natural Resource Products and Uses: An Example

A product mix table for a southwestern, ponderosa pine forest is shown below. The values represent estimates of the level of natural resource production and use for alternative management practices.

The comparisons presented in this table show accurate estimates of gains and losses in natural resource products and uses for four management alternatives. The M_o management alternative estimates production and use values when management practices are not changed. M_1 are values for conversion of moist forest sites to grassland, and M_2 and M_3 represent values resulting from uneven-aged and even-aged management practices, respectively. Examining the results of the M_0 alternative, the annual outputs per acre will be 5.9 inches of water, forage for 0.150 pounds of livestock, and 150 cubic feet of timber growth, and no timber will be cut. If M_0 is selected as the best alternative after assessing the gains and losses in natural resources production and use, then existing management practice should be continued.

Product mix table of four alternative management practices for a southwestern ponderosa pine forest

	Management alternatives			
Items (per acre)	M_0 As is	M_1 Convert	M_2 Uneven-aged	M_3 Even-aged
Water (in.)	5.9	8.7	6.3	7.1
Livestock (lbs gain)	0.150	1.05	0.000993	0.594
Timber cut (ft^3)	0	318	170	135
Timber growth (ft^3)	150	88.5	195	185

BENEFITS AND COSTS

Determining the benefits, or values, associated with alternative natural resources management practices and the costs of implementing and maintaining these management practices are prerequisite to selecting the best management of natural resources in a multiple-use, ecosystem-based context. These benefits and costs must be identified at a common point in the sequence of economic activities (e.g., obtaining the raw materials from the ecosystem, processing the raw materials into usable products or uses, marketing the products or uses, etc.) associated with the management practice. Placing values on benefits at one point and costs at another point leads to incorrect evaluation and incorrect selection of the management practice.

Benefits

Benefits of natural resources management practices can be easily estimated in some instances. For example, the value of harvested timber for processing into wood products is often quantified in the marketplace. However, calculations based on the commercial market for only one wood product (lumber) become obsolete if markets for other wood products (pulpwood or particleboard) become available. The presence of these additional market outlets can increase the expected monetary returns from harvesting timber by making a previously unwanted wood product marketable.

Values for forage, wildlife, and other natural resources are often estimated by proxy values because marketplaces for trading these primary resources are lacking. Values of forage may be estimated from the economic returns of weight gains in livestock. Wildlife values may be estimated from the anticipated revenue generated from hunting (e.g., license fees) or viewing wildlife resources (e.g., park entrance fees). Values from outdoor recreation are often inputted by knowing the levels of expenditure for identified activities.

Values of some natural resources are almost impossible to estimate. A true value for water is one example, because there is not only one marketplace for water that is flowing from a watershed. Nevertheless, attempts should be made to place a value on water and other natural resources lacking an easily defined marketplace. Otherwise, conclusions obtained from an economic analysis of alternative management practices could be misleading.

Costs

Information on the cost of implementing and sustaining a variety of natural resources management practices is available in the literature. Unfortunately, much of this information reflects the cost for a particular economic situation at a specific point-in-time; consequently, it may not be easy to adjust costs to reflect current conditions. Costs expressed in terms of *physical input-output variables,* which characterize a management practice and land area in terms of time-independent, benefit-cost data, can help overcome this problem (Box 13.2).

The flexibility of time-independent, benefit-cost information allows reevaluation of a natural resource management practice through time as local economic conditions change. A management practice that was initially considered impractical could subsequently become operational and economically feasible, for instance, when increased market outlets for natural resource products become available or when there is a change in wage or machine cost rates.

Economic Evaluations

Economic objectives are often used for evaluating the feasibility of increasing the number of products or uses derived from an alternative natural resources management practice, for example, water production, livestock, or wood production. These economic objectives can include:

- Maximizing benefits.
- Maximizing returns on an investment.
- Achieving a specified production goal at the lowest possible cost.

Box 13.2

Physical Input-Output Variables: Some Examples

Physical inputs to natural resources management practices include time allotted for supervision, labor, equipment, and cost of materials. Outputs specify the total work completed, for example, the number of acres planted, cleared, or otherwise treated, the number of trees planted, or the pounds of forage seed sown. Costs are determined by multiplying these inputs by current wage rates, machine operating rates, and material costs. Summation of all of the costs divided by the number of work units (acres planted, cleared, or treated, number of trees planted, etc.) is an estimate of the *average unit-cost* for a management practice.

Information satisfying the first two objectives include: estimates of the physical responses in products and uses of the natural resource that result from each alternative management practice; values for products and uses; and costs of implementing and maintaining each management practice considered. To satisfy the third objective, production goals, such as attaining a specified level of timber production at the lowest possible cost, are generally derived through decision-making processes.

Economic evaluations of natural resources management practices can consist of an array of economic analyses—each of which is designed to help people make better decisions. An individual economic analysis may provide a one-answer solution to the problem of selecting a management practice that maximizes the collective benefits to an area. However, a group of economic analyses based on different criteria results in an array of items for decision makers to consider. Such an array may include:

- Estimates of multiple-use production and use that are associated with each alternative management practice.
- Cost estimates for the alternative management practices.
- Low cost solutions for the different goals of multiple-use production and use.
- Gross and net benefits associated with the range of each alternative management practice.
- Investment returns and benefit-cost ratios associated with each alternative management practice.

Methods of making economic evaluations of natural resources management practices are found in Gregersen et al. (1987),

Loomis (1993), Boardman et al. (1996), Loomis and Walsh (1997), and Niemi and Whitelaw (1997).

DECISION MAKING

The problems that arise with the integration of different natural resources management practices are often complex and involve several management disciplines. For example, harvesting timber impacts forage production, wildlife habitats, soil and water conservation, and in all likelihood an array of other natural resource products and uses. The impacts of harvesting timber may be beneficial to some of these natural resources, while they are negative to others. Furthermore, the proponents of forest industries, environmental groups, and political organizations tend to interpret the effects of timber harvesting from different and often conflicting viewpoints. Therefore, a natural resource manager is faced with a decision-making task—a task of obtaining fair and equitable solutions to the biophysical, socioeconomic, and political problems of harvesting timber. A similar decision-making task confronts the manager implementing almost any form of natural resources management.

Fundamentals of Decision Making

The decision-making process in natural resources management is generally viewed in four sequential steps:

- Problem recognition.
- Specification of strategies.
- Specification of the decision criterion or criteria.
- Selection of the optimum strategy.

Problem recognition is the most important step, since unrecognized problems will almost certainly remain unsolved. If problems remain unrecognized, resources devoted to solving unimportant problems or problems no longer requiring attention are not available for solving important problems.

Each alternative to solve a recognized problem is a *strategy.* Decision making is largely a matter of selecting one of the strategies available. However, it is pointless to consider alternatives that can not possibly be implemented. On the other hand, omitting feasible alternatives from consideration can adversely affect the final decision, for example, if one or more of the omitted strategies is superior to any of the strategies being considered.

A *decision criterion* is a measure that evaluates each strategy and expresses the desirability of the outcomes obtained from each strategy. Selection of the appropriate decision criterion is only possible if a natural resources manager (the decision maker) has clearly identified the management objectives. The relationships among the potential decision criteria and management objectives must also be properly understood. In profit-oriented organizations, decision criteria are usually some measure of economic desirability (e.g., discounted present value, rate of return, management costs). An appropriate decision criterion may not be available in situations involving goods, services, or amenities that can not be readily valued in economic markets. This problem is often confronted in natural resources management that involves water, rangeland resources, wildlife, and recreation considerations.

Once the decision criterion (criteria) has been specified, values of the decision criterion are calculated for all of the strategies being considered. The strategies can then be ranked in order of desirability according to the decision criterion values (Box 13.3). Large values indicate high desirability with some criteria (discounted present value, rate of return), while small values indicate high desirability for other criteria (cost).

Decision-Making Process

There are many decision-making processes, or techniques, available to natural resources managers. A linear programming formulation, which has a single objective, is a relatively simple decision-making model. For example, a natural resources manager may wish to minimize the cost of timber harvesting. This is a linear objective because two linearly related variables are involved: timber production and the cost of timber harvesting. The linear assumption of the simple decision-making model is inappropriate for more complex natural resources management problems, however.

Some problems confronted by natural resources managers involve a number of objectives, for instance, minimizing the costs of a management practice while at the same time maximizing the benefits obtained from forage production, wildlife habitats, and soil and water conservation. A set of such objective functions will likely be subject to a number of linear and nonlinear constraints. Problems of this more complex but more realistic form of decision making can be analyzed with models that are based on multiple criteria (objective) decision-making (MCDM) methods. MCDM techniques are designed to find the preferred solutions to problems, where the discrete alternatives are evaluated against acceptance criteria (factors) ranging from quantitative (costs) to qualitative (desirability).

Box 13.3

Desirability of Strategies: An Exception

Knowledgeable natural resources managers do not always select the strategy with the most desirable decision-criterion value. For example, when computing the decision-criterion value it would be irrational to prefer a higher ranked strategy (A) that is based on questionable information and data rather than a lower ranked strategy (B) that is based on reliable information. In this case, the argument can be made that the natural resources manager adopted a new (revised) decision criterion in choosing strategy (B) over strategy (A). However, it can be difficult for the manager to quantify this new criterion to the satisfaction of all decision makers.

Descriptions of commonly used decision-making methods, requirements for their use, and their applications in natural resources management are found in Dress and Field (1987), Kent and Davis (1988), Anderson et al. (1994), Yakowitz and Hipel (1997), and El-Swaify and Yakowitz (1998).

SUMMARY

More responsive and efficient use is made of the natural resources on a site by recognizing potential products and uses, and then combining these potentials into integrated natural resources management programs. The emphasis on integrating natural resources management into multiple-use, ecosystem-based management practices is becoming increasingly important as society continues to demand more benefits and uses from natural resources.

After completing this chapter, you should be able to:

- Explain why the management of water, forage, wood, and other natural resource products and uses should be planned in the framework of a broad, multiple-use, ecosystem-based concept.
- Discuss how natural resources management can be integrated into the multiple-use, ecosystem-based concept by considering the levels of natural resource products and uses, associated benefits and costs, and economic evaluations of natural resources management practices.
- Understand the importance of economic evaluations in selecting natural resources management practices from a set of alternatives.
- Discuss the role of decision support systems in integrated natural resources management.

REFERENCES

Anderson, D. R., D. J. Sweeney, and T. A. Williams. 1994. An introduction to management science: Quantitative approaches to decision making. West Publishing Company, St. Paul, Minnesota.

Boardman, A. E., D. H. Greensberg, A. R. Vining, and D. L. Weimer. 1996. Cost-benefit analysis: Concepts and practice. Prentice-Hall, Upper Saddick River, New Jersey.

Dress, P. E., and R. C. Field, editors. 1987. The 1985 symposium on systems analysis of forest resources. Georgia Center of Continuing Education, Athens, Georgia.

Dykstra, D. P. 1984. Mathematical programming for natural resource management. McGraw-Hill Book Company, New York.

El-Swaify, S. A., and D. S. Yakowitz, editors. 1998. Multiple objective decision making for land, water, and environmental management. Lewis Publishers, Boca Raton, Florida.

Gregersen, H. M., K. N. Brooks, J. A. Dixon, and L. S. Hamilton. 1987. Guidelines for economic appraisal of watershed management projects. FAO Conservation Guide 16, Rome, Italy.

Kent, B. M., and L. S. Davis, technical coordinators. 1988. The 1988 symposium on systems analysis in forest resources. USDA Forest Service, General Technical Report RM-161.

Loomis, J. B. 1993. Integrated public lands management: Principles and applications to natural forests, parks, wildlife refuges, and BLM lands. Columbia Publishing Company, New York.

Loomis, J. B., and R. G. Walsh. 1997. Recreation economic decisions: Comparing benefits and costs. Venture Publishing, Inc., State College, Pennsylvania.

Niemi, E., and E. Whitelaw. 1997. Assessing economic tradeoffs in forest management. USDA Forest Service, General Technical Report PNW-GTR-403.

Ridd, M. K. 1965. Area-oriented multiple use analysis. USDA Forest Service, Research Paper INT-21.

Worley, D. P., G. L. Mundell, and R. M. Williamson. 1965. Gross job time studies—an efficient method for analyzing forestry costs. USDA Forest Service, Research Note RM-54.

Yakowitz, D. S., and K. W. Hipel. 1997. Multiple objective decision making for "Lokahi" (balance) in environmental management. Applied Mathematics and Computation 83:97–115.

14

Importance of Research to Natural Resources Management

SCIENCE IS knowledge attained through study (research) and practice (management). Therefore, researchers and managers are both scientists. It is important that scientists commit to sharing their scientific knowledge with all people. Unfortunately, researchers, managers, decision makers, and interested lay people are occasionally at odds on the direction that natural resources managers should take. Much of this disagreement can be traced to differences in environmental philosophy.

ROLES OF RESEARCHERS AND MANAGERS

The role of researchers and managers in the management and use of natural resources to satisfy human desires has been articulated nicely by Stoltenberg et al. (1970) among others. Managers assist individuals and society by identifying and clarifying objectives for how natural resources should be managed. Managers also identify alternative approaches to achieving objectives; help evaluate and compare these alternatives; and assist in selecting the most promising opportunities for achieving stated objectives. However, managers can spend a disproportionate amount of time supervising activities to implement objectives and have insufficient time to devote to problem-solving efforts. This situation is unfortunate, because natural resources managers often make their most valuable contribution by helping people make decisions.

Natural resources managers' contributions to society are generally measured by how effectively they help people satisfy their needs. The value of researchers to society is largely determined by how much their efforts increase a manager's efficiency. Research has three general purposes. First, research develops new concepts, tools, practices, products, and services for managers. Second research answers questions that arise when a management process is implemented. And third, research answers questions that arise while research is conducted. All too often basic questions must be answered satisfactorily before the first two purposes of research can be addressed efficiently.

THE RESEARCH CONTINUUM

Research can be viewed as a continuum. Natural resources managers are at the beginning of the spectrum. Managers are concerned principally with people's problems. When managers lack the information needed to help evaluate and compare alternative solutions to a problem, they often experiment with alternative approaches until a satisfactory solution is obtained. In this process many managers become surrogate researchers.

Next on the continuum are researchers who attempt to answer managers' immediate questions on resource-use potentials, management technologies, trends, and values. Following these researchers are developmental researchers, who create new alternatives that will hopefully help managers solve both immediate problems and anticipated problems. Researchers who serve as a clientele of other researchers are next on the continuum. These researchers usually come from the disciplines of biology, chemistry, and physics. They provide facts, describe relationships, and provide other inputs needed by those scientists who conduct more applied developmental research. The ultimate client for researchers and managers is the general public.

Successful problem anticipation and problem-solving become more difficult as the distance on the spectrum between researchers and managers increases. While researchers may focus on their immediate clients, a large portion of their activities is usually intended to help solve management-related problems. Research planners need to anticipate these problems accurately and articulate their contents appropriately. Information necessary to solve problems comes from managers, environmentalists, and those members of society who are stakeholders in an enterprise. Information comes from all of the sources on the research continuum, confirming that research and management should be viewed as a mutually interconnected process.

MEETING PEOPLE'S NEEDS

Researchers and managers, working together, face changing expectations on how natural resources should be managed and used to meet people's needs. At the beginning of the twenty-first century, the managerial strategies, scientific knowledge, and technology necessary for obtaining the traditional uses of natural resources only partially satisfy people's interests and demands. Therefore, researchers and managers must be responsive to the different viewpoints that people have regarding resource use. Such a perspective embraces a land stewardship that balances protecting natural environments and sustaining the products and services needed by people.

TOOLS OF RESEARCH

The appendices at the end of this book provide information to natural resources managers on the tools employed in research and the analysis of research investigations. This information includes overviews on the use of plot studies to help evaluate the effectiveness of natural resources management practices (Appendix 2); applications of statistical methods to estimate population parameters and test hypotheses about these parameters (Appendix 3); applications of computer simulation models in natural resources management (Appendix 4); and applications of geographical information systems to store, retrieve, and update information for natural resources management (Appendix 5). More information on these research tools is found in the references presented in the respective appendices.

SUMMARY

Natural resources managers should recognize the important role of research as a source of information needed for better management practices. At the same time, however, researchers should appreciate the informational needs of natural resources managers. Only by understanding their respective roles can researchers and managers work together to satisfy people's expectations for natural resource products and uses.

At the end of this chapter, you should be able to:

- Describe the continuum of research efforts that lead to the implementation of better natural resources management practices.
- Explain why researchers and managers should work together in meeting people's needs relative to natural resources management.
- Discuss the applications of the tools of research in natural resources management.

REFERENCES

Kessler, W. B., H. Salwasser, C. W. Cartwright, and J. A. Caplan. 1992. New perspectives for sustainable natural resources management. Ecological Applications 2:221–225.

National Research Council. 1990. Forestry research: A mandate for change. National Academy of Science, Washington, D.C.

Stoltenberg, C. H., K. D. Ware, R. J. Marty, and J. D. Wellons. 1970. Planning research for resource decisions. Iowa State University Press, Ames, Iowa.

Appendix 1

English to Metric Conversions

Quantity	English unit	Metric unit	To convert English to metric multiply by
Length	inches (in.)	centimeters (cm)	2.54
	inches (in.)	millimeters (mm)	25.4
	feet (ft)	meters (m)	0.304
Area	square inches ($in.^2$)	square millimeters (mm^2)	645.16
	square feet (ft^2)	square meters (m^2)	0.0929
	acres	hectares (ha)	0.405
	square miles (mi^2)	square kilometers (km^2)	2.59
Volume	cubic inches ($in.^3$)	cubic centimeters (cm^3)	16.4
	cubic feet (ft^3)	cubic meters (m^3)	0.0283
	gallons	liters (L)	3.79
Velocity	miles per hour (mi/hr)	kilometers per hour (km/hr)	1.61
	feet per second (ft/sec)	meters per second (m/sec)	0.305
Flow	cubic feet per second (ft^3/sec)	cubic meters per second (m^3/sec)	0.0283
	gallons per minute (gpm)	liters per second (L/sec)	0.0631
Mass	ounces [avdp] (oz)	grams (g)	28.3
	pounds [avdp] (lb)	kilograms (kg)	0.454
Density	pounds per cubic foot (lb/ft^3)	grams per cubic centimeter (g/cm^3)	0.0160
	pounds per cubic foot (lb/ft^3)	kilograms per cubic meter (kg/m^3)	16.0
Temperature	degrees Fahrenheit (°F)	degrees Celsius (°C)	(5/9)(°F − 32)
Special	number per acre (no/acre)	number per hectare (no/ha)	2.47
	square feet per acre ($ft^2/acre$)	square meters per hectare (m^2/ha)	0.229
	cubic feet per acre ($ft^3/acre$)	cubic meter per hectare (m^3/ha)	0.709
	pounds per acre (lb/acre)	kilograms per hectare (kg/ha)	1.12

Appendix 2

Plot Studies

A FIRST STEP in evaluating the feasibility of implementing a natural resources management practice may involve the use of plot studies to determine the effectiveness of the practice in satisfying its stated purpose. Cause-and-effect information can often be obtained through the use of small, homogeneous plots that are representative of controlled conditions. Plot studies are utilized in seedling survival tests, water-balance investigations, precipitation-runoff evaluations, and analyses of the effects of vegetative management practices on water regimes, forage and timber production, and wildlife habitats. Results from plot studies can indicate a need to follow up on a larger scale, such as landscape-level experiments.

Plot studies also help to furnish data for the development and testing of mathematical relationships that are needed as components of computer simulation models (see Appendix 4).

PLOT CHARACTERISTICS

Plots can range from one or two square feet to several acres in size. Measurements obtained from smaller plots often may not represent the natural conditions compared to those obtained from larger plots, although the use of smaller plots can be sufficient in reconnaissance studies. Larger plots frequently yield more reliable results for operational extrapolation. Boundary distortions imposed on ecosystem processes sampled by plots can be a problem. Increasing the plot size often helps to reduce these effects. Another possible solution to this problem is establishing plots with no artificial boundaries.

Circular plots are easier to install in the field than rectangular plots of the same size. A center-point needs to be established with a circular plot. The specified radius is inscribed on tape that is inserted at the center-point of the plot. A circular plot having the same size of a rectangular plot has less perimeter. As a consequence, decisions relating to the measurement of borderline conditions are minimized.

The plot size and shape to use in a study is determined in relation to the objectives of the study, the local field conditions encountered, the ease of establishment, and if necessary, maintenance. The expense and types of materials required for installation are also factors to consider.

EXPERIMENTAL DESIGNS

Many experimental designs are available for consideration when planning plot studies. For example, to study the rates of runoff and soil loss a researcher may want to experiment with varying combinations of vegetative treatments, soil types, or soil conditioning methods. Considering the large number of experimental designs available to study these combinations, it is important that the plot studies be designed to provide clear-cut answers to the questions being asked.

Experiments are always subject to variability. The selected experimental design should allow for control of the most extraneous sources of variation. Plot studies should also be designed so that known sources of variation are deliberately varied over as wide a range as is necessary. Plot studies need to be designed so that their variability can be eliminated from the estimate of chance variations. Experimental designs achieving this purpose include *completely randomized designs, randomized block designs, Latin square designs, and factorial designs*. Descriptions of these and other experimental designs, including analyses of the measurements obtained from them, are found in the references at the end of this appendix.

Procedures for data collection and those responsible for collecting data must be clearly specified at the onset of a plot study.

Data collection methods need to be unbiased and consistent throughout the study; otherwise, interpretations of the data can be misleading. A data-management system can be helpful in storing and retrieving the data sets obtained from plot studies for display and analysis.

OTHER OBSERVATIONS

Plot studies are attractive to researchers because they provide an opportunity for the study of small, homogeneous systems; plot studies allow for control of inputs into the study and they provide accurate measures of outputs. Plot studies are easier to establish and maintain and less expensive to conduct than larger landscape-level studies. For example, artificial barriers can be constructed to constrict water flow or prevent leakage, which enables the water budget components to be measured more accurately. Comparisons of the effects of several land-use treatments on different vegetation types can be made on a small-scale with a series of plots.

A limitation of plot studies can be the inability to extrapolate the results obtained to larger, more complex landscape-level ecosystems. For example, the rate of soil moisture depletion determined from a plot study cannot be used to determine soil moisture decreases on the entire watershed system. Plot and landscape-level studies provide the same types of information or information for the same purposes. However, plot studies are designed to improve the knowledge of specified ecological processes, allowing interpretation and better understanding of the responses of larger ecosystems to natural resources management practices. Landscape-level studies provide the information necessary to estimate the responses of the larger ecosystems to implementation of the management practices.

Another drawback to plot studies can be the high costs of replicating the research in the system in which the results will be applied. Large-scale application of research results from plot studies requires wide-scale replication. The inherent variability of the ecosystem in question could require monitoring plots for ten years or more. Long-term monitoring may be required to attain a required level of precision and to meet desired extrapolation standards.

REFERENCES

Green, R. H. 1979. Sampling design and statistical methods for environmental biologists. John Wiley & Sons, Inc., New York.

Lentner, M., and T. Bishop. 1986. Experimental design and analysis. Valley Book Company, Blacksburg, Virginia.

Natrella, M. G. 1963. Experimental statistics. Handbook 91, U.S. National Bureau of Standards, Washington, D.C.

Ostle, B. 1963. Statistics in research: Basic concepts and techniques for research workers. Iowa State University Press, Ames, Iowa.

Snedecor, G. W., and W. Cochran. 1967. Statistical methods. Iowa State University Press, Ames, Iowa.

Appendix 3

Statistical Methods

STATISTICS IS the science of analyzing collected measurements of a population and inferring "truths" about the measurements from the results of the analysis. There are two general objectives of statistics—to estimate parameters that describe populations and to test hypotheses about the parameters. An example of the first objective is the estimation of the mean (average) and variability of measurements around the mean (variance) of a population. Having satisfied this objective, a statistician's next task is to prescribe proper sampling techniques for collecting the required source data and then to calculate the desired statistics. An example of the second objective is testing the hypothesis that the estimated population mean equals or exceeds a value representing a theoretical expectation. It is the statistician's job in this case to select the appropriate tests to evaluate the hypothesis.

SAMPLING TECHNIQUES

It can seem desirable to count all the individuals (census) in a population. However, a census is rarely possible and a sample is usually taken. The sample size, size and shape of individual sampling plots (see Appendix 2), and sampling design must be considered before a population is sampled.

Sample Size

A primary objective of sampling is to take enough measurements to obtain a desired level of precision, no more, no less. Deciding on the number of samples to take depends on the inherent variability of the population being sampled and the desired level of precision, or the *allowable error*. An estimate of the population variance is necessary to calculate the sample size. Sample size is defined as n, the units (plots, points, etc.) in a sample. This estimate can be obtained from a preliminary sample of the population. A t-value at the specified level of probability is also required, and the level of precision must be specified. Formulas for calculating the sample size for different sampling designs are found in the references at the end of this appendix.

Sampling Designs

Samples are measured to estimate desired population parameters. The problem is deciding which sampling design is most efficient in terms of the sampling objectives and inherent characteristics of the population. Many sampling designs and variations of sampling designs exist. However, the basic sampling designs are simple random-sampling, stratified random-sampling, and systematic sampling.

- The fundamental idea behind *simple random-sampling* is that every possible combination of the n units sampled has an equal chance of being selected. The selection of any given individual sampling unit is also independent of the selection of all other units. Previous knowledge of the distribution of a population is often used to increase the precision of a sample.
- *Stratified random-sampling* takes advantage of available information about a population by grouping homogeneous units of the population together, based on some inherent characteristic (vegetative cover, soil parent material, slope-aspect combinations, etc.). Each homogeneous unit, called a *stratum,* is then sampled by employing a simple random-sampling design, and the estimates for each stratum are combined to estimate (overall) population parameters.
- Individual sampling units are allocated according to a predetermined pattern in *systematic sampling*. Systematic sampling has been and continues to be used for two reasons: locating individual sampling units in the field is often easier and cheaper than is the case with other sampling designs, and there is often an intu-

ition that a sample that is deliberately spread over a population is more representative than a simple, random sample. While statisticians may accept the first reason, they are less willing to accept the second. The estimation of sampling errors that are associated with a systematic sample requires more knowledge about the population being sampled than is usually available.

BASIC STATISTICS

A measure of the central value of the measurements obtained from a sample of a population is usually required. The *mean* is the most familiar and commonly used measure of central tendency. Another measure is the *median*. The median for a series of values obtained from a sample of a population and arranged (ranked) in order of size, is the value that has an equal number of items on each side. Another measure of central tendency is the *mode*. When a series of random values are arranged by classes and frequencies (a frequency distribution), one class (or a few classes) will generally represent the highest frequency of occurrence. The class (or classes) with the highest frequency of occurrence is the mode (or modes).

Measures of dispersion indicate the extent to which individual observations in a series vary from the measure of central tendency. The *range* is the total interval between the smallest and the largest values in a series of random values. The range provides preliminary information on the variability of random values in a series. A more useful measure of dispersion is the *variance*, which describes whether the individual, random values in a series are close to the estimated population mean or are spread out. The larger the variance, the more scattered the values are from the mean. The square root of the variance is the *standard deviation*, which is used in calculating the *coefficient of variation*. The coefficient of variation, the ratio between the standard deviation and the mean, allows variance to be compared across populations, each having a different mean.

The *standard error of the mean* is a measure of dispersion among a set of sample means. However, it is not necessary to obtain a set of simple, random samples to calculate the standard error of the mean. A satisfactory estimate can be obtained from the

Table APP3.1. Formulas for some basic statistics

Mean $= \bar{x} = \frac{\Sigma X}{n}$
Variance $= s^2 = \frac{\Sigma(X - \bar{x})^2}{n - 1} = \frac{\Sigma X^2 - \frac{(\Sigma X)^2}{n}}{n - 1}$
Standard deviation $= \sqrt{s^2}$
Coefficient of variation $= \frac{s}{\bar{x}}$
Standard error of the mean $= s_{\bar{x}} = \sqrt{\frac{s^2}{n}} = \frac{s}{\sqrt{n}}$
Confidence limits (internal) $=$ Cl $= \bar{x} \pm ts_{\bar{x}}$

source data of one simple, random sample. The reliability of an estimated population parameter is estimated by *confidence limits*. Confidence limits are a function of the standard error of the mean and a *t*-value selected to meet the level of precision specified. While confidence limits can be structured around many statistics, they are most commonly established about an estimated population mean.

Formulas for some of the more commonly used basic statistics that are calculated from data sets obtained in simple random-sampling are presented in Table APP 3.1.

HYPOTHESIS TESTING

It is often important to know whether the estimated means of two or more groups of sample data are different. The groups of sample data compared can represent different treatments, for example, the effects of different types of vegetative cover on the rate of soil loss on a watershed.

Two hypotheses are typically evaluated in hypothesis testing. The *null hypothesis* states that there is no statistical difference between the means of two or more groups of sample data; the *alternative hypothesis* states that there is a statistical difference. The results of tests of hypothesis allow statements to made relating to the acceptance or rejection of the null hypothesis. If the null hypothesis is accepted, the alternative hypothesis is rejected and vice versa.

A *t*-test is commonly used to compare two groups of sample data, while an analysis of variance is used to compare two or more groups of sample data. All tests of hypothesis require the following:

- A valid sample has been made.
- The variables measured are normally distributed (see below).
- All groups of sample data have the same population variance.

Procedures for testing hypotheses are outlined in certain references listed at the end of this appendix.

REGRESSION ANALYSIS

A statistical tool that is frequently used in analyzing measurements is *regression analysis*. Regression analysis requires the selection of appropriate mathematical models to quantify the relationships between a *dependent variable* (for example, streamflow discharge) and one or more *independent variables* (rainfall amount, antecentent soil moisture, etc.). These mathematical models are approximations based on sample data; therefore, they are subject to sampling variations.

Simple Regression

A *simple regression* defines a relationship between a dependent variable (Y) and an independent variable (X). A simple *linear regression* uses a straight-line to show the relationship between two variables. The simple linear regression model is:

$$Y = a + bX \tag{A3.1}$$

where Y = dependent variable
X = independent variable
a, b = regression constants

A *nonlinear* response is often found between two variables in natural resources studies. When this occurs, the relationship between the variables must be transformed into a linear form to simplify the analysis. For example, suspended-sediment concentration

(Y) versus streamflow discharge or tree height versus tree diameter are typically nonlinear responses that are log-transformed. This model is represented by:

$$Y = aX^b \quad \text{(A3.2)}$$

transformed to:

$$\log Y = \log a + b(\log X) \quad \text{(A3.3)}$$

Other nonlinear simple regression analyses can involve parabolic, exponential, and other nonlinear relationships, which require a transformation of the variables.

Multiple Regression

A dependent variable can be related to more than one independent variable. For example, peak- or low-flow streamflow discharges from a watershed (Y) are likely to be related to some combination of annual rainfall (X_1), watershed area (X_2), and the percent of area in impoundments and wetlands (X_3). If this relationship can be estimated by a *multiple regression analysis*, more precise estimations of peak- or low-flow discharge (the dependent variable) are obtained than is possible by a simple regression. A multiple regression model is:

$$Y = a + b_1X_1 + b_2X_2 + \ldots + b_nX_n \quad \text{(A3.4)}$$

where a = regression constant

$b_1, b_2, \ldots, b_n$ = slope of the relationships between respective independent variables $X_1, X_2, \ldots, X_n$ with the dependent variable Y

FREQUENCY ANALYSIS

A *frequency analysis* of a data set is performed to determine the probability of occurrence of a specified event (e.g., an annual rainfall amount on a watershed, level of forage production on a rangeland, or diameter growth of trees in a forest stand) of a stated magnitude (e.g., inches of rainfall, pounds of forage per acre, or inches in tree diameter). The pattern of frequency occurrence of the units in each of a series of equal classes is a *frequency distribution function.*

By knowing the frequency distribution function, it is possible to determine what proportions of the individuals in the population are within specific size classes. Each data set representing a population has its own unique frequency distribution function. However, there are certain distribution functions that are often used in analyses of natural resources data sets, including the normal, binomial, and Poisson distributions.

A *normal frequency distribution,* which is the familiar bell-shaped distribution (Figure APP3.1), is used widely in the statisti-

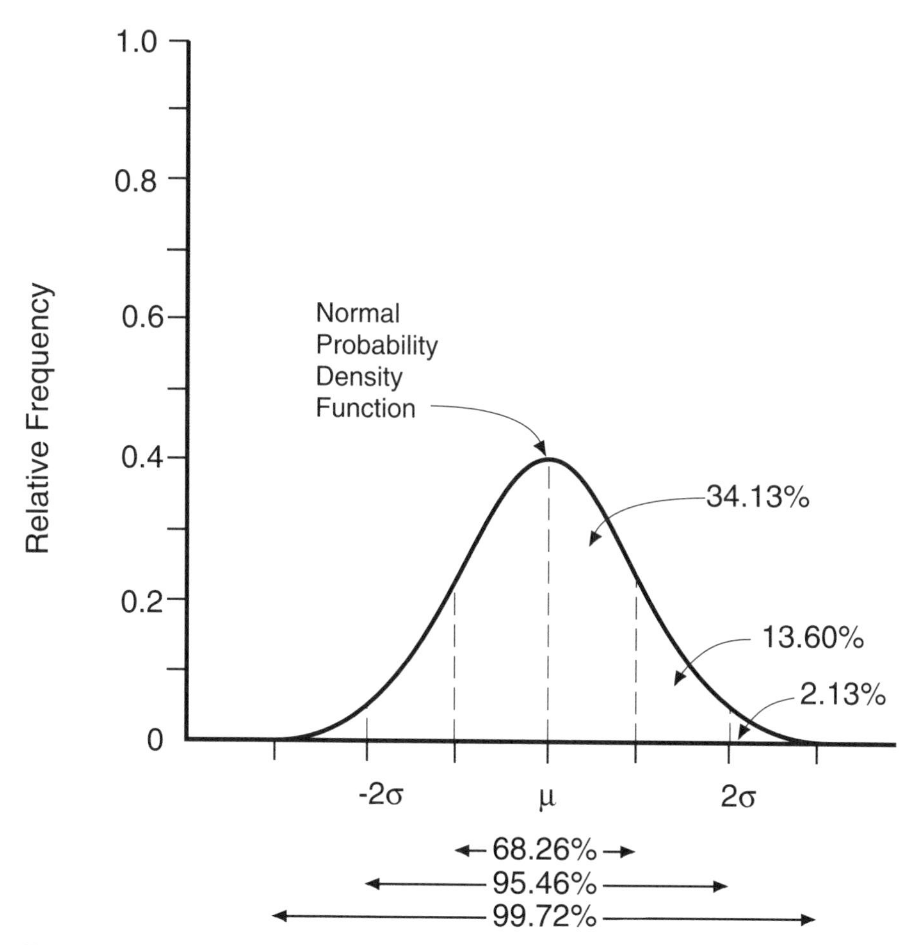

Figure APP3.1. The bell-shaped normal frequency distribution. Percentages of items lie within the indicated limits.

cal analyses of natural resources measurements. In a normal frequency distribution:

- The mean, median, and mode are identical in value.
- Small variations from the mean occur more frequently than large variations from the mean.
- Positive and negative variations about the mean occur with equal frequency.

Properties of the *binomial frequency distribution* and the asymmetric *Poisson frequency distribution* are found in references on statistical theory. One of the tasks of a statistician is identifying the appropriate function for a situation, and then applying the proper statistical calculations to estimate the needed population parameters.

OTHER OBSERVATIONS

Points that natural resources managers should keep in mind when applying and interpreting statistics are:

- Understand the statistical method applied—Statistics is a complex subject. Statisticians themselves are often unsure of their own conclusions and commonly discuss them with their colleagues for verification. To minimize misuse of the method, be sure that the statistical method used is thoroughly understood.
- Keep it simple—Where there is a choice of statistical methods, the simpler one should be applied. There is little need to apply the latest, detailed analytical tool if an older, simpler method provides the manager with the information required. There is also less chance of the analysis "going wrong" when the simpler method is used.
- Plan ahead—The sampling design and analysis procedure should be planned before the outset of a study. Unfortunately, a common practice is leaving the statistical analysis and interpretation to a time when it is impossible to do an adequate job.
- Adopt a critical attitude when examining the results of a statistical analysis. Try to let the source data, rather than preconceptions, lead the interpretation process.
- A statistical analysis must not be reduced to a routine process. A

natural resources manager should not be led astray by computer analysis packages. While computer programs are tools for expediting analyses, remember *computers do not analyze*. Remain familiar with useful statistical programs but do not surrender the analyses to them.

- Respect professional judgment—Training and experience gives a manager the ability to "see" the elements and relationships in natural resources management better than a layman.

REFERENCES

Conover, W. J. 1971. Practical non-parametric statistics. John Wiley & Sons, Inc., New York.

Freese, F. 1964. Linear regression methods for forest research. USDA Forest Service, Research Paper FPL-17.

Freese, F. 1967. Elementary statistical methods for foresters. USDA Forest Service, Agriculture Handbook 317.

Green, R. H. 1979. Sampling design and statistical methods for environmental biologists. John Wiley & Sons, Inc., New York.

Hoshmand, A. R. 1998. Statistical methods for environmental and agricultural sciences. Lewis Publishers, Boca Raton, Florida.

Larson, H. 1982. Introduction to probability and statistical inference. John Wiley & Sons, Inc., New York.

Lentner, M., and T. Bishop. 1986. Experimental design and analysis. Valley Book Company, Blacksburg, Virginia.

Milton, J. S. 1992. Statistical methods in the biological and heath sciences. McGraw-Hill, Inc., New York.

Natrella, M. G. 1963. Experimental statistics. Handbook 91, U.S. National Bureau of Standards, Washington, D.C.

Snedecor, G. W., and W. Cochran. 1967. Statistical methods. Iowa State University Press, Ames, Iowa.

Thomas, R. 1976. Numerical nonsense: A discussion of commonly unrecognized problems in collecting, analyzing, and drawing conclusions from data. USDA Forest Service, Region Five, San Francisco, California.

Tukey, J. 1977. Exploratory data analysis. Addison-Wesley, Reading, Massachusetts.

Williams, E. J. 1959. Regression analysis. John Wiley & Sons, Inc., New York.

Appendix 4

Computer Simulation Models

COMPUTER SIMULATION models are representations of "real-world systems" that help explain and predict how ecosystems respond to natural resources management practices. Using models can give a better understanding of the impacts of management practices. These models should be based on a *systems approach* to simulation to be effective. A *system*, for example, an ecosystem, is an interacting assemblage of elements that function together for the same purpose. Connections among the parts that constitute the system being modeled are emphasized in the systems approach.

DEVELOPMENT

Computer simulation models are often composites of mathematical relationships, some empirical, some cause-and-effect, and some based on theory. Attempts to explain or predict the impacts of natural resources management practices on increasingly complex systems require more detail and complexity as the model is formulated. However, an understanding of ecosystem functioning is usually not sufficient to represent every process mathematically. This can lead to a need to develop a computer simulation model that is calibrated to fit parameters and relationships to the conditions encountered. The *calibration process* involves adjusting parameters until the computed response approximates the observed response. Once calibrated, the computer simulation model can then be used to estimate the response of ecosystems to natural resources management practices.

Methods of Development

Computer simulation models are developed in many ways. One approach is through regression analysis (see Appendix 3), where the natural relationship between selected variables is expressed in a curve-fitting process. Knowing the behavior of variables allows for the selection of one specific regression model over another. However, selecting the most appropriate regression model for a particular data set is somewhat of an art. The statistical properties of various regression models should be understood before a choice is made. When cause-and-effect relationships cannot be adequately identified, empirical relationships can be derived.

The assemblage of one or more predictive functions (such as those developed by regression analyses) can simulate the functioning of a natural ecosystem. The response of an ecosystem to different inputs and input levels is simulated with such models. The sensitivity of model outputs, or the response to changing the value of regression constants (parameters) in the model, can be determined. Furthermore, requirements for additional support data can often be determined.

Developing a computer simulation model includes assembling mathematical equations, which represent the predictive functions of an ecosystem, in a flowchart diagram (Figure APP4.1). The functions and linkages that are identified in the flowchart are then translated into a set of instructions written in a computer language. The model components expressed in this language are then entered into a computer along with the appropriate descriptive data and any other required input data. Finally, outputs that predict the response of the system to specified inputs are obtained by executing the model.

Many computer simulation models operate by using a simulation exercise where input data is entered as the answers to questions posed to the user by a computer program. These simulators are *interactive models*, in contrast to assemblages of large data records on magnetic disks or tapes used for input to *batch models*.

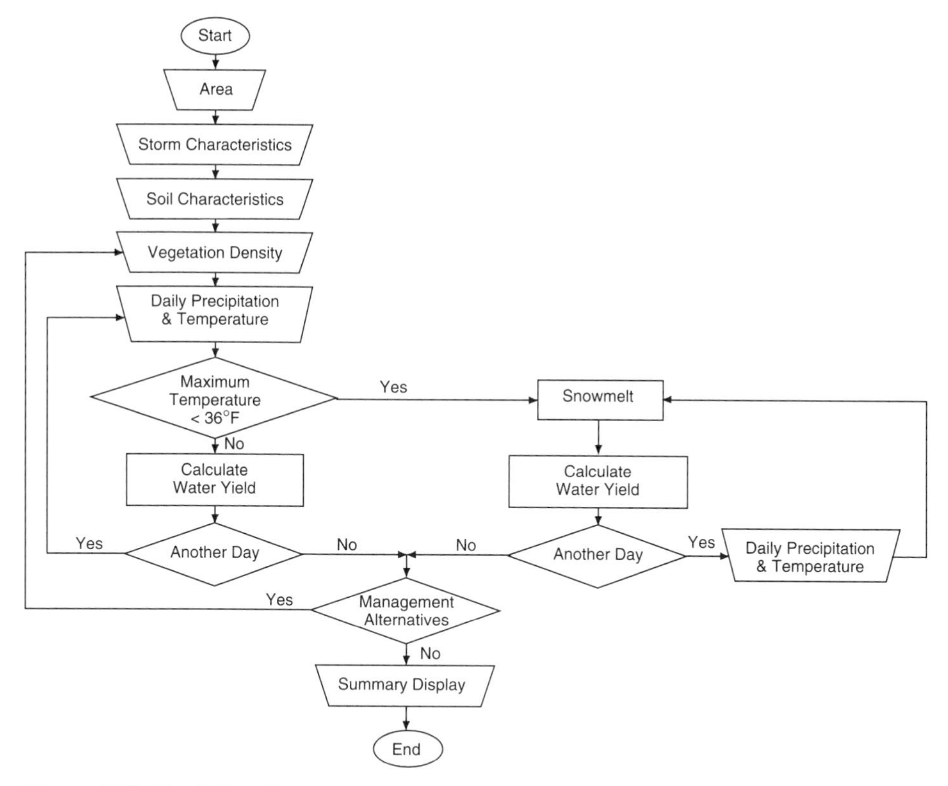

Figure APP4.1. A flowchart diagram, with descriptive text being used to represent mathematical equations, for a computer model simulating the water yields from a watershed. One component of the model simulates water yields from rainfall events (left) and the other component simulates water yields from snowmelt events (right).

[A description of the formulation and application of the computer simulation model illustrated by the flow chart is found in Ffolliott and Guertin 1988.]

Interactive models, also called *user-friendly models*, are often used to learn more about the possible changes in ecosystem functioning that may be caused by alternative natural resources management practices.

Desirable Characteristics

A computer model that simulates the effects of management practices on natural resources should have the following characteristics:

- Accurate representation of the management practices included in the proposed model and the ability to simulate the effects of these management practices.
- Inclusion of time-dependent phenomena, because the effects of most management practices can be expected to change over time.
- The ability to accept inputs that describe the spatial variability of both the ecosystem landscape and the management practices on the landscape. The spatial distribution of the effects of management practices generally has an influence on the effects of these management practices.
- The ability to operate on readily available databases. Computer simulation models should not include data requirements that are difficult, costly, or time-consuming to collect or acquire.

APPLICATIONS

A question frequently asked of a natural resources manager is: What are the anticipated responses of natural resources to a proposed management practice? Applications of computer simulation models can be useful in answering this question. Models developed to estimate the natural resource response to management practices include the following:

- Changes in streamflow, peak flows and low flows and changes in the physical, chemical, and bacteriological quality of surface water.
- Composition, production, and utilization of forage plants and livestock carrying capacities.
- Growth, yield, and quality of standing timber.
- Habitat quality for wildlife and fish species.
- Quality of visual resources.

Models that simulate the responses of natural resources to management practices should include the following characteristics:

- Accuracy of prediction—It is desirable that the development of computer simulation models includes knowing the error statistics. Models having minimum bias and error variance are generally desired.

- Simplicity—Simplicity refers to the number of parameters estimated and the ease with which the model can be explained to potential users.
- Consistency of parameter estimates—This is an important consideration in the development of models that use parameters estimated by optimization techniques. Computer simulation models are likely to be unreliable if the optimum values of the parameters are sensitive to the duration of the data record (file) representing the parameters used in the simulation exercise or if the values vary widely between similar ecosystems.
- Sensitivity of results to changes in parameter values—It is not desirable for a model to be sensitive to input variables that are difficult to measure and costly to obtain.

SELECTION CRITERIA

A large number of computer simulation models are available to help a manager predict the effects of a management practice on a specified natural resource. However, these models are likely to vary in complexity, input data and other operational requirements, and the type of information they can provide. A problem confronting the manager is selecting the most appropriate model from the available choices. A computer simulation model is helpful only if it meets the following criteria:

- The intended purpose for the model is clearly defined. This includes knowing the required spatial and temporal resolutions, which helps to define the type of data needed and level of accuracy for collection and input of information.
- Input data and other information needed to operate the model is available.
- The model is suitable for the physical, biological, and social conditions of the site to be modeled.
- The time and economic constraints for operating the model are known.
- The appropriate hardware systems and supporting software programs are available.
- The user has the appropriate skills, experience, and background for creating and using the model.

Once a computer simulation model has been selected, it must be calibrated and tested to determine its appropriateness for the situation where it will be applied. Even if a model has been successfully applied to conditions similar to those being considered, the model is likely to still require some testing because each site is unique.

OTHER OBSERVATIONS

Selecting computer simulation models for a specific application often involves a compromise between theory or completeness and practical considerations. Locally derived models are generally based on limited data and are often empirical in nature, limiting their application elsewhere. More complete and theoretical simulation models may be better suited for widespread application, but they can require input data or other information that are not available for local applications.

A computer simulation model may need to be developed when it is decided that available models are not suitable to meet particular needs. Model development requires knowing the necessary functional relationships of the processes being modeled. These relationships can be based on theory, plot studies, landscape-level experiments, or other previous research. The developmental process can seem circular because computer simulation models are often used when the results from plot studies are lacking. But information from plot studies can form a basis for model developmental in situations where appropriate models are not available and must be constructed.

REFERENCES

Ek, A. R., S. R. Shifley, and T. E. Burk, editors. 1988. Forest growth and modelling prediction. USDA Forest Service, General Technical Report NC-120.

Feldman, A. D. 1981. Models for water resources simulation: Theory and experience. Advanced Hydrology 12:297–418.

Ffolliott, P. F., and D. P. Guertin. 1988. YIELD II: An interactive computer model to simulate water yields from southwestern ecosystems. In: Modeling agricultural, forest, and rangeland hydrology: Proceedings of the 1988 international Symposium. American Society of Agricultural Engineers, St. Joseph, Michigan, pp. 72–78.

Hann, C. T., H. P. Johnson, and D. L. Braaakensiek. 1982. Hydrologic modeling of small watersheds. Monograph Number 5, American Society of Agricultural Engineers, St. Joseph, Michigan.

Larson, C. L. 1971. Using hydrologic models to predict the effects of watershed modifications. In: National symposium on watersheds in transition. Colorado State University, Fort Collins, Colorado, pp. 113–117.

Kirkby, M. J., P. S. Naden, T. P. Burt, and D. P. Butcher. 1993. Computer simulation in physical geography. John Wiley & Sons, Inc., New York.

Roberts, N., D. Andersen, R. Deal, M. Garet, and W. Shaffer. 1983. Introduction to computer simulation: The systems dynamics approach. Addison, Wesley Publishing Company, Reading, Massachusetts.

Wiant, H. F., D. O. Yandle, and W. E. Kidd. 1986. Forestry microcomputer software symposium. Division of Forestry, West Virginia University, Morgantown, West Virginia.

Appendix 5

Geographic Information Systems

THERE HAS been an unparalleled evolution of computing technologies in recent years. Geographic Information Systems (GIS) are examples of this accelerating development. GIS is becoming an important tool in the planning and implementation of management practices on natural ecosystems. These systems are largely designed to satisfy the geographical information needs of managers and researchers. GIS is optimized to store, retrieve, and update geographic information about ecosystems. They are programmed to process spatial information upon demand in a format that meets the users' informational needs.

FORMATS

A set of geographically registered data layers is maintained in a GIS for retrieval and analysis (Figure APP5.1). These layers can be stored in either *raster* or *vector* formats.

Raster Format

The data layer in a raster (cell-based) system is represented by an array of rectangular or square cells, each of which has an assigned value. Advantages of a raster format include:

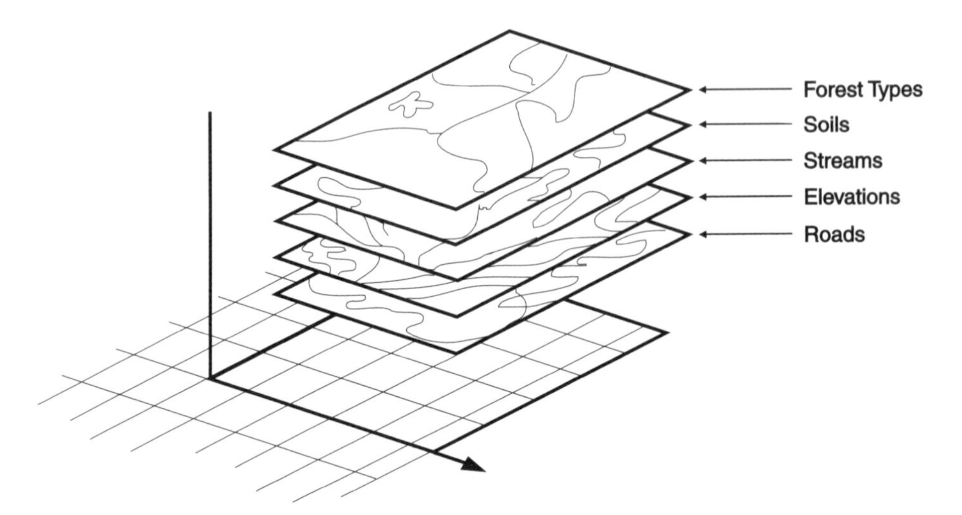

Figure APP5.1. A GIS is conceptualized as a set of geographically registered data layers (adapted from van Roessel 1986).

- The geographic location of each cell is implied by its position in the cell matrix. The matrix can be stored in a corresponding array in a computer, provided sufficient disc space is available. Therefore, each cell can be easily addressed in the computer according to its geographic location.
- The geographic coordinates of the cells do not require storage, since the geographic location is implied in a cell's position.
- Neighboring locations are represented by neighboring cells. Therefore, neighboring relationships can be conveniently analyzed.
- Discrete and continuous data sets are accommodated equally well, which facilitates the intermixing of the two data types.
- Processing-algorithms are simpler and easier to write than is the case in vector systems.
- Map unit boundaries are presented by different values. When these values change, the implied boundaries change.

Disadvantages of a raster format are:

- Storage requirements are larger than the requirements for vector systems.
- The cell size determines the resolution at which the resource is represented. It is difficult to adequately represent linear features.

- Image access is often sequential, meaning that a user may have to process an entire map to change a single cell.
- Processing descriptive data are more cumbersome than is the case with a vector system.
- Input data are mostly digitized in form. Therefore, a user must execute a vector-to-raster transformation to convert digitized data into a form appropriate for storage.
- It can be difficult to construct output maps.

Vector Format

The line work on a map is presented by a set of connected points in a vector (line-based) system. A line segment between two points is considered a vector. Coordinates of the points are explicitly stored, and the level of connection between points is implied through the organization of the points in the database. Advantages of a vector format are:

- Much less storage space is required than is needed for a raster system.
- Original maps can be represented at their original resolutions.
- Forests, streams, roads, and other resource features can be retrieved and processed individually.
- It is easier to associate a variety of resource data with a specified resource feature.
- Digitized maps do not need to be converted to a raster form.
- Stored data can be processed into raster-type maps without a raster-to-vector conversion.

Disadvantages of a vector format include:

- Locations of the vertex points need to be stored explicitly.
- The relationship of these points must be formalized in a topological structure, which can be difficult to understand and manipulate.
- Algorithms for accomplishing functions that are the equivalent of those implemented on a raster system are more complex. Furthermore, the implementation can be unreliable.
- Vectors cannot represent continuously varying spatial data. A conversion to raster is required to process these types of data.

GIS APPLICATIONS

GIS, often used together with remote sensing techniques and computer simulation models, has a number of applications for natural resources managers and researchers. Applications of GIS include:

- Inventorying and monitoring—GIS and remote sensing can be used to keep inventory information current. Resource inventories can include available quantities, location, and information on whether resource availability is growing, shrinking, or being maintained at some level. Monitoring more than inventorying is a major new thrust for natural resource managers. Monitoring is crucial for two reasons:
  - Management agencies often make agreements after purposeful planning processes, environmental impact decisions, and court-ordered decrees require the use of specified management practices to monitor impacts.
  - Managers often need monitoring information to make changes in ineffective management practices.
- Management planning—GIS allows for management at a large spatial scale, which is valuable in planning for the management of large ownerships or for more than a single owner. For example, forestry managers would no longer need to deal only with one forest stand at a time. They could place the management of the stand into the context of the ecosystem. Management planning that is not ecosystem-based can result in unwanted and unexpected downsides, such as cumulative effects or ecosystem fragmentation.
- Policy setting—Science is currently an important focus in setting natural resource policy. GIS can be used to prepare and display information on management alternatives for review by officials making policy decisions. For example, in the early 1990s the policy for the protection of late-succession Pacific Northwest forest ecosystems and their associated species was being decided. This policy-setting process may have been quicker and more efficient if the management alternatives considered by the Scientific Panel and Forest Ecosystem Management Assessment Team (FEMAT) had been better mapped and displayed (Sample 1994). Many of the resource maps presented to FEMAT were hand-drawn, which

is a tedious, labor-intensive process that is prone to transcription errors.

- Research—Researchers experience more problems when conducting a series of replicated, landscape-level field studies than they do for smaller scale studies. With GIS, researchers can address these problems by synthesizing resource data, developing concepts, and displaying findings. Information about patches, edges, connectivity, cumulative effects, and dispersed and aggregated activities can be included in the process.
- Decision making—With GIS it is possible to have interactive collaborations among managers, policy setters, researchers, and other stakeholders in the decision-making process. The time when managers and researchers prepare and present management alternatives to the public without incorporating current science, social perspectives, and economic interests has passed. Today, professionals are participants and facilitators in the decision-making process, and GIS technology has played a significant role in this development.

SOURCES OF ERROR

The spatial data sets in a GIS are obtained from maps, aerial photographs, satellite imagery, and traditional and global positioning surveys. Each source of information involves a number of transformation-steps from the original measurements to the final digital coordinates. Each step can also introduce an error into the system. The origin of common errors in a GIS include:

- Field measurements—All positional information ultimately relies on field measurements. These measurements can be relatively precise (such as those needed to define legal property boundaries) or they can be approximate (such as eyeball locations of inventory plots on topographic maps). Global Positioning Systems (GPS), which consist of a control segment, a constellation of satellites, and GPS receivers, can provide positional information at lower costs and higher resolution throughout than traditional surveying. However, GPS is only appropriate for a limited number of data layers.
- Maps—Manual or automated map digitization is currently the most common form of spatial data entry; therefore, it has the

greatest impact on the accuracy of digitized data sets. Manual digitizing involves overlaying a piece of transparent paper or Mylar on a map that has been placed on a digitizing surface and then tracing the features of interest. The accuracy of the original map used is a determinant of spatial data accuracy regardless of the digitizing method used.

- Imagery—Imagery is a common source of natural resource spatial data for both initial database development and updates. Most imagery comes from aerial cameras and satellite scanners, although video cameras and airborne scanners are also becoming popular. For fifty years or more, aerial photographs have been routinely used in resource mapping. Among the factors affecting the accuracy of photo-derived data layers are tilt and terrain distortion and lens or camera distortion. Automated classification of satellite imagery is a method of land-cover mapping. Classification converts multiband reflectance data into a single-layer land cover map, which is then registered to a geographic coordinate system. Therefore, the accuracy of class-boundary location is a function of classification accuracy, geometry of the image, and quality of the registration.
- Digitization—Positional accuracy during digitization is affected by the equipment used and the skills of the operator. Currently used digitizers possess accuracy levels (and precision) of better than 0.001-inch (0.0025-centimeter). However, errors of varying magnitudes (up to 15 percent) can still be observed for digitized arcs and polygons.
- Coordinate registration—Errors can occur when converting from digitizer coordinates to the coordinate system of the map projection used for printing the source map. Positional errors can occur at any of the steps in the process, including identifying control points in both the geographic and digitizer space; choosing a mathematical transformation and estimating the coefficients; and applying the transformation to the digitized data in producing the output layer. While large blunders are easily detected, small or random errors are not. Control points must be obtained from ground surveys or, when field measurements are lacking, control is commonly digitized from geographic coordinate points drafted on the source map (for example, Universal Transverse Mercator (UTM) graticule intersections drafted on 1:24,000-scale base maps).

OTHER CONSIDERATIONS

Other considerations for the use of GIS as a tool in natural resources management are data organization, database functions, input, query and analysis, display, reporting, user interface, and hardware. Data organization addresses the raster versus vector issue. Database functions cover topics such as operating system use. Input includes digitizing and any input needed from external sources. Overlay consideration is a query and analysis topic. Display relates mainly to graphic outputs, while reporting considers the preparation of tabular reports. User interface is concerned with menus versus command modes and other methods of control. To an observer of the system, hardware is the most prominent consideration, but it is part of the environment for the experienced user.

REFERENCES

Bernhardsen, T. 1999. Geographic information systems: An introduction. John Wiley & Sons, Inc., New York.

Bolstad, P. V., and J. L. Smith. 1992. Errors in GIS. Journal of Forestry 90(11):21–29.

Clarke, K. 1997. Getting started with geographic information systems. Prentice Hall, Upper Saddle River, New Jersey.

Congalton, R. G., and K. Green. 1992. The ABCs of GIS. Journal of Forestry 90(11):13–20.

Longley, P. A., M. Goodchild, D. Maguire, and D. W. Rhind, editors. 1998. Geographic information systems: Principles, techniques, applications and management. John Wiley & Sons, Inc., New York.

McNulty, S. G., J. M. Vose, W. T. Swank, J. D. Aber, and C. A. Federer. 1994. Landscape scale forest modeling: Data base development, model predictions and validation using a geographic information system. Climate Research 4:223–231.

Quattrochi, D. A., and M. F. Goodchild, editors. 1997. Scale in remote sensing and GIS. Lewis Publishers, New York.

Sample, V. A., editor. 1994. Remote sensing and GIS in ecosystem management. Island Press, Washington, D.C.

Star, J., and J. Estes. 1990. Geographic information systems: An introduction. Prentice Hall, Englewood Cliffs, New Jersey.

van Roessel, J. W. 1986. Guidelines for forestry information processing. FAO Forestry Paper 74, Rome, Italy.

Index